Environmental Security Concept and Dimensions

ENVIRONMENTAL SECURITY CONCEPT AND DIMENSIONS

Narottam Gaan

Publications

KALPAZ PUBLICATIONS
DELHI - 110052

Environmental Security Concept and Dimensions

Rs. 750

ISBN : 81-7835-295-8

Published in 2004 in India by
Kalpaz Publication,
C-30, Satyawati Nagar,
Delhi-110 052
Email : kalpaz@hotmail.com

Lasertype Setting by: Tact Computer Graphics, Delhi
Printed at: Mehra Offset Press, Delhi

Dedicated to My Revered Gurudev

**Let the tiny lotus bud of my love dangling
tenderly in the cosmic breeze of Maya
held firmly on the timeless stalk of Thy Love
blossom into an eternal flower of Infinite Beatitude
at the very touch of illuminous
unfaded shine of All Resplendent Light
of my Gurudev, whom millions of stars, moons
and suns fail to express.**

Dedicated to My Revered Gurudev

Let the tiny lotus bud of my love swaying

tenderly in the cosmic breeze of Maya

held firmly on the timeless stalk of True Love

blossom into an eternal flower of Infinite Beatitude

at the very touch of illuminous

unbidden shine of All-Resplendent Love

of my Gurudev, whom millions of stars, moons

and suns fail to express.

Contents

Preface

It was my privilege to provide a Foreword to Dr. Narottam Gaan's groundbreaking *Environment and National Security: the Case of South Asia* (New Delhi: South Asian Publishers, 2000). Now, several years later, I have with this Preface the opportunity to update and expand upon that earlier brief contribution of mine. At this time I examine the overall (general) concept of environmental security as it has evolved and matured in my mind since my first serious consideration of the matter over two decades ago. I do so in hopes that it will offer a conceptual basis for, and complement to, Dr Gaan's masterful in-depth analysis which follows herein.

Here I emphasize that environmental security is inextricably linked with social security, the two together comprising the necessary expansion of the traditional concept of national security. I define environmental security as a tripartite combination of: (a) the sane exploitation of our natural resources, including the sustainable use of the renewable ones; (b) the elimination of medically unacceptable air and water pollution; and (c) the protection, especially through nature reserves, of the global biodiversity and ecosystem functioning.

The central responsibility of the sovereign state (that is, of its government) to provide its citizens with national security is established, but is nonetheless suggested to be unachievable without: (a) substantial cooperation with neighbouring states as well as with the world community of states; and (b) active cooperation and input by the private sector, including nongovernmental organizations (NGOs), corporations, including transnational (multinational) corporations (TNCs), academia, philanthropic organizations, and the public at large. I thus introduce the entire subject by an examination of the nature of state sovereignty and the forms of state-to-state interactions,-both international (governmental) through treaties, etc: and transnational (non-governmental) through commerce, trade, etc. This leads us into a state's external responsibilities, that is, its obligations to the other states and to the extra-territorial (common-heritage) domains. Next I address a state's internal responsibilities, that is, its obligations to its citizenry, which have been expanding in recent decades from an emphasis on military security to the far broader and more comprehensive social/environmental notion of security being emphasized here.

The presently deteriorating environmental security throughout much of the world becomes the jumping off point for my presenting means of countering such deterioration through a combination of mutually supportive actions by the state's public and private sectors. Here I point out the hitherto insufficiently recognized role of the business sector. It is suggested that the existing paucity of effective environmental constraints on TNCs throughout much of the world should (until such time as this international shortcoming is rectified by effective regulatory mechanisms) be countered especially by a combination of NGO pressures supported by public awareness, and self-imposed constraints emanating from within the business sector itself. Finally, I stress the overall need for bringing into balance the available natural resources with the global human population as an ineluctable prerequisite for achieving and maintaining comprehensive human security, as are the need for society to accept nonviolent means of conflict resolution and the need for society to function under the rule of law, both domestically and internationally, within a framework of justice and equity. To emphasize the necessity to avoid armed conflict, I am pleased to be able to refer here to my 'Environmental and Ecological Consequences of War, in Dr Gaan's forthcoming book *Environmental Issues and Challenges: A Critique on Intemational Relations Theories.*

The Notion of State Sovereignty

The two Treaties of Westphalia (Westphalen) of 1648 not only ended the Thirty Years' War, but also laid the foundation for the international system of sovereign states within which — for better or worse — the world now tries to function. Indeed, just since the end of World War II, the number of generally recognized sovereign states has increased from 53 to the current 192 or so. These 192 intensely sovereign entities divide up virtually all of the earth's land area (and a substantial fraction of the ocean as well) — and with it all the people who inhabit the earth, plus an appreciable fraction of its natural resources. This seemingly haphazard patchwork quilt of independent political units is the product of an incredibly varied and tortuous social history, one that derives whatever legitimacy it claims to enjoy from three and a half centuries of global precedent and acceptance.

Fortunately, most of the numerous sovereign states of the world abide each others' existence to some greater or lesser extent, and thus for the most part interact peacefully and (they must assume) to their own advantage. The states accomplish such generally peaceful coexistence through a host of formal and informal mechanisms, so that the very real potential for anarchy and violence is in fact held in check in most parts of the world for much of the time. Two basic sorts of global interaction exist: (a) those between or among the sovereign states themselves, that is, from government to government — the international (interstate) interactions; and (b) those among the peoples of the world more or less distinctly outside governmental channels — the transnational interactions. These two forms of interaction are variously — and increasingly — linked to each other: one example of such linkage is the slowly growing influence of NGOs over the pursuit and outcome of interstate interactions; another example is the regulatory control that governments exercise, with varying degrees of success, over TNCs.

External State Obligations

Interstate relations function primarily through the mechanism of international treaties, of which there are many thousands, variously bilateral, regional, or universal in scope. These contractual agreements have in turn set up a plethora of bilateral, regional, and global governance regimes, many of which function through concomitantly established area-specific or subject-specific intergovernmental agencies (administered by permanent secretariats), but which for the most part have only ineffectual compliance mechanisms. Important universal environrfiental examples include the: (a) 1971 Wetland (Ramsar) Convention [UNTS 14583]; (b) 1972 World Cultural and Natural Heritage Convention [UNTS 15511]; (c) 1973 Convention on Trade in Endangered Species (CITES) [UNTS 14537]; (d) 1979 Migratory Animal Convention [UNTS 28395]; (e) 1982 Law of the Sea [UNTS 31363]; and (f) 1992 Biological Diversity Convention [UNTS

30619]. Important regional examples include the: (a) 1979 European Long-range Air Pollution Convention [UNTS 21623]; (b) 1979 European Wildlife Conservation Convention [UNTS 21159]; and (c) 1982 Mediterranean Protected Areas Protocol [UNTS 24079].

Regrettably, however, the closest that the global family of sovereign states comes to a functioning world government (its closest approach to world order) is through the fundamentally weak United Nations system — kept ineffectual by its lack of a judicial system which is accepted as being compulsory and unconditional (one that should be applicable on such basis both to states and to individuals); and kept unjust by the veto I power of the 'Big Five' in the United Nations Security Council.

A sovereign state has some basic external obligations. The two most important of such obligations is to respect the independence and the equality of all the other sovereign states. (Of course, each treaty that a state enters into is done to provide some advantage to it, but which at the same time diminishes somewhat its absolute sovereignty.) A further obligation that derives from the basic ones (and has in fact become explicitly enshrined in a number of international instruments) is that a state carry out no activity that would cause harm to another state, or even to domains beyond national jurisdiction (that is, to such extraterritorial domains as the high seas or the global atmosphere). Finally, I should here mention a more nebulous (regrettably, some might even say dubious) principle of external state obligation to look upon the various extra-territorial domains as being the common concern, or — even more strongly — as being the common heritage of humankind. The slowly emerging notion of a common heritage of humankind implies that both the benefits and the management burdens of such extra-territorial domains require some sort of equitable apportionment among the states and peoples of the world.

The Concept of National Security

Having outlined the notion of a sovereign state and its obligations to the other sovereign states with which it shares the world, it is now time to discuss the internal obligations of each state, those toward its citizenry, past, present, and future. Especially since the end of World War II, it has been ever more often articulated that a state's obligations to its citizenry go well beyond traditional military security (either individual or collective).

As the most conspicuous early example of the articulation of such sentiment, the United Nations-sponsored 1948 Universal Declaration of Human Rights [UNGA Res 217A (III)] proclaimed a series of such rights prominent among them that: (a) everyone has the right to lite, liberty, and security of person; (b) all are entitled to equal protection of the law; (c) everyone has the right to take part in the government of his or her country; (d) everyone has the right to a standard of living adequate for health and well-being, explicitly including a right to food, clothing, housing, medical care, and unemployment and old-age security; and (e) everyone has the right to education.

Although not spelled out in so many words by the aspirational or hortatory (that is, non-binding) *Universal Declaration of Human_Rights*, a declared human right automatically translates into an inferred state obligation. Indeed, most of these human rights subsequently became formally accepted by large numbers of states by virtue of their becoming parties to two complementary (and partially overlapping) treaties — the 1966 International Covenant on Economic, Social and Cultural Rights [UNTS 14531] and the 1966 International Covenant on Civil and Political Rights [UNTS 14668]. Moreover, both of these widely accepted treaties explicitly recognized that the rights they promulgated entailed concomitant internal state obligations. Various national constitutions also guarantee such rights to their citizens, as do a number of more recent (though sometimes less widely adopted) human-rights treaties. Thus, it can be seen that as such human rights and civil liberties — as a package comprising the notion of social security — become ever more widely enshrined, they concomitantly expand the obligations of a state, thereby automatically also expanding the notion of national security.

The next quantum leap to have been taken regarding rights and obligations has derived from a slowly emerging recognition that the global environment (the global biosphere) had to be conserved to some significant extent in order for the basic human rights being proclaimed ever more widely to in fact become realizable. This breakthrough in human awareness found its first important glimmers in the widely endorsed hortatory *Declaration* of the 1972 United Nations Conference on the Human Environment (held in Stockholm) [UNGA Doc A/CONF.48/14/Rev.l], which included among its principles that: (a) everyone has the fundamental right to an environment of a quality that permits a life of dignity and well-being; (b) the natural resources of the earth must be safeguarded for the benefit of present and future generations; (c) the capacity of the earth to produce renewable resources must be maintained; and (d) the benefits of the non-renewable resources are to be shared by everyone. Note here that all of these principles were strictly human-oriented (anthropocentric; homocentric), that is, the environment was to be coddled strictly for the benefit of humankind.

Then came another hortatory United Nations proclamation, the 1982 *World Charter for Nature* [UNGA Res 37/7], a *Charter* reiterating and strengthening much of the 1972 *Declaration*, but innovative in two important ways: (a) by demanding a respect for nature partly in its own right — that is, its proclaimed justifications were both human-oriented and nature-oriented (ecocentric; biocentric); and (b) by holding both states and-individuals responsible for its realization (the latter especially, a remarkable departure for the United Nations). Indeed, the nature-oriented justifications for biodiversity protection suggest an additional category of state responsibility, to wit, an obligation toward all non-human life in its own right.

Beyond these (and a number of other) international hortatory expressions of human obligations towards the environment, the first major binding (treaty-based) set of relevant environmental obligations derives from the very widely adopted 1992 Biological Diversity Convention [UNTS 30619] (a major outcome of the 1992 United Nations Conference on Environment and Development, held in Rio de Janeiro). This treaty has among its objectives the conservation of biodiversity (that is, of living organisms) and the sustainable use of its components for humankind. Importantly, it goes on to codify the responsibility of each contracting state not to cause environmental damage beyond its borders.

It is thus clear that national security is becoming ever more widely recognized as being a far broader concept than its traditional limitation to military security. Indeed, if one gives some thought to the more recently established array of human rights comprising social security, national security can now be seen as an inextricably interconnected two-tier affair, with the attainment of environmental security providing the inexorably required underpinning to offer the opportunity to achieve the full panoply of social security; but at the same time, with environmental security being unattainable without the achievement of social security.

Environmental Security

With the increasingly recognized broad environmental/social approach to national security having been explicated, it will now be valuable to review for a moment what exactly is meant by environmental security. In fact, to achieve environmental security, three more or less interlocking elements must all be satisfied:

- First, the natural resources of the earth—both territorial and extraterritorial; and whether extractive or non-extractive — must be utilized in a sane manner
- In that regard, frugality, efficiency, and recycling are the keys to the sane utilization of the non-renewable resources, whereas sustainable harvesting and sustainable waste disposal are the keys to the sane utilization of the renewable resources. The renewable resources being utilized as a source of harvestable crops include forestlands, rangelands, marine fisheries, croplands, and aquifers; the renewable resources being utilized as disposal sinks include freshwater courses, the ocean, the atmosphere, and landfills.
- Second, medically unacceptable pollution, especially of air and water, must be eliminated. It might be added here that since human tolerance levels to noxious wastes are for the most part narrower than those for most wildlife, the needed public health measures will be serving a salutary dual purpose.
- Third, in addition to avoiding unnecessary damage to the flora and fauna on that huge fraction of the world we have arrogated to ourselves for municipal, industrial, agricultural, transportational, and other needs, it is essential that some relatively modest fraction of each of the major habitat types throughout the world be specifically set aside and kept largely free of human intrusion purely for purposes of nature protection, that is, purely for conserving the biodiversity they support. Such establishment of protected natural areas must be done for both long-term human-oriented and nature-oriented reasons, the former including a huge array of crucial ecosystem services; and the latter including the intrinsic rights that must be assigned to all living things.

Achieving Environmental Security

Having established what comprises environmental security, having suggested that environmental security is a human right and inextricable from social security — and for that reason at least a state obligation to provide — the questions remain of the extent to which this responsibility is actually being met and the means by which any identified shortfalls could be eliminated.

Indeed, the world as a whole, and very few if any states by themselves, have achieved the mandated and ever so necessary environmental security. Indicators of a lack of environmental security are strikingly easy to find, among them especially: (a) the increasingly unsustainable exploitation of the world's forestlands and marine fisheries; (b) the increasingly unsustainable exploitation of the atmosphere for the disposal of gaseous wastes (prominent among those the so-called greenhouse gases); (c) the increasing insufficiencies of sanitary water; (d) the ever increasing flows of environmental refugees (both internal and cross-border); and (e) the ever increasing numbers of species extinctions.

In the earlier discussion I pointed to the responsibility of each state to individually provide environmental security for its citizens. However, reference to the numerous treaty-based group endeavors — whether bilateral, regional, or universal in scope — certainly indicates a widespread recognition that meeting many of the existing environmental challenges is simply beyond the scope and abilities of any single state, whether for financial, competency, motivational, or other reasons. One obvious additional reason for such individual inability is that political boundaries and environmental (ecosystem) boundaries seldom coincide, thereby necessitating the cooperative regional or even global approaches to which I have already alluded.

It is thus sadly clear that the great majority of states — that is, their governments — have to date failed to provide their citizenry with the necessary environmental security. But there are two basic components that comprise the society of a state: its public sector and its private sector. Although a state's public sector of governance, administration, and judiciary has a central role to play in providing its citizens with environmental security, this can only be achieved with the active cooperation and continuing input of the private sector, including especially its educational, religious, service, philanthropic, NGO, and business components. Indeed, with the private business sector representing such a huge and influential fraction of society, it becomes particularly important to examine its diverse contributions to environmental security, both positive and negative, and how these might all be channeled toward beneficial outcomes.

The business sector — broadly defined here to include commerce, finance, industry, insurance, and trade — constitutes a major segment of the private sector of society not only within a state, but also transnationally. Domestic legislation to establish and enforce environmental responsibilities for this sector within the borders of the two dozen or so wealthy industrialized states of the world is more or less adequate—although, of course, there is always room for improvement. Moreover, the democratic institutions in most of those states permit their citizens to exercise some surveillance functions, thereby helping to insure compliance with the regulatory legislation. However, such is not generally the case within the remaining 165 or more states. Indeed, some of those states do not adopt constraining environmental (or social) safeguards precisely as a means of attracting TNCs to their shores. As a result, TNCs are frequently subject to only minimal environmental or social accountability to governmental authorities in their often far-flung operations, a most disturbing situation when it comes to some of the really large or otherwise powerful ones. And, to their shame, the business sector actually spends huge amounts of money annually to prevent or weaken environmental legislation by a variety of direct and indirect means.

Nonetheless, countervailing forces, both public and private, are beginning to nudge the business sector in the right direction — among other approaches, pointing out the goodwill to be created, the investor risk reductions to be achieved, and the overall economic benefits (both short-term and long-term) to be derived from various environment-friendly operations. Some of the more important of those forces are international environmental NGOs, consumer groups, research institutes, academicians, and even TNC shareholders.

A key intergovernmental initiative was that of the 1992 United Nations Conference on Environment and Development, *inter alia*, in its *Agenda 21* calling upon the business sector to strengthen its support of sustainable development. As to nongovernmental initiatives, one can, *inter alia*, point to the Coalition for Environmentally Responsible Economies (Boston, USA; 1989—), which has been able to convince several dozen or more corporations to expand their annual reports to include environmental sustainability accounting, in accordance with some rather detailed guidelines it has developed. The Rocky Mountain Institute (Snowmass, USA; 1982—) and the Wuppertal Institute for Climate, Environment and Energy (Wuppertal, Germany; 1989—) are persuasive about the economic benefits to be gained from adopting environmentally sound practices; and the Institute of Social and Ethical Accountability (London, UK; 1996–) offers ethical justifications and guidelines for social (including environmental) accounting. The NGO, Eco-Tex (Cologne, Germany; 1992–) monitors the textile industry (production, manufacture, and trade) from an environmental standpoint and offers 'green' certification. And Greenpeace (Amsterdam, Netherlands; 1970–) has become well known for the vigorous publicizing of what it considers to be grievous corporate environmental transgressions.

It is also heartening to be able to highlight some initiatives emanating from within the business sector itself. For example, ethical codes of conduct covering both social and environmental issues are being voluntarily adopted by a growing number of TNCs and other businesses. Adoption of such codes has been in many instances abetted by the International Chamber of Commerce (Paris, France; 1972—) and by the World Business Council for Sustainable Development (Geneva, Switzerland; 1990—). Among relevant business coalitions one can mention some that were established by environmental NGOs, for example: (a) the Center for Environmental Leadership in Business, founded by Conservation International (Washington, USA; 1987—); and (b) the Business Environmental Leadership Council, founded by the Pew Center for Global Climate Change (Arlington, USA; 1998—). Among those established on the initiative of the business sector itself should be mentioned: (a) the International Petroleum Industry Environmental Conservation Association (London, UK; 1974—); and (b) e-mission 55—Business for Climate (Bonn, Germany; 2001 —). And, in face, a number of TNCs seem to now be moving toward sustainable practices of 'eco-efficiency'.

On the other hand, whereas growing numbers of major TNCs are publicly embracing social values and principles of environmental sustainability, we must not lose sight of the fact that many of their actions seem to remain sharply at odds with their newly professed credos. The so-called triple bottom line — a combination of economic, social, and environmental performance — increasingly espoused by corporations still appears to remain widely ignored in practice, with profit-making often still remaining the unambiguous nom. The world of international banking (both public and private) has also not as yet really adjusted to the environmental imperatives.

Maintaining Environmental Security: Towards a Sustainable Society

To solidify any gains in environmental security, all states should be asked to publish an annual 'green' accounting of their progress toward environmental sustainability (as has been pioneered by the Netherlands). Moreover, those intergovernmental agencies which facilitate international trade — especially the World Trade Organization (Geneva, Switzerland; 1995—), the North American Free Trade Agreement (Washington, USA; 1994—), and the European Free Trade Agreement (Geneva, Switzerland; 1960—)—must incorporate overriding ('bottom-line') environmental protection constraints into their rules and regulations. Such intergovernmental agencies as the International Bank for Reconstruction and Development (the World Bank) (Washington, USA; 1945—), the International Monetary Fund (Washington, USA; 1945—), and the Organisation for Economic Co-operation and Development (Paris, France; 1961—) must do likewise. Not only must all of these intergovernmental agencies become less subservient to free-trade and development pressures— and more clearly 'green' — but at the same time more open ('transparent') in their deliberations. The World Trade Organization has even gone so far as to rebuff attempts by the United Nations Environment Programme to participate in its deliberations in an observer capacity.

To maintain environmental security it will, among other approaches, be crucial to solidify whatever gains have been made in the business sector, which accounts for such an immense fraction of all human activity. Governments, intergovernmental agencies, NGOs, business associations, and the public at large can together accomplish this through a combination of strengthened regulations (applicable to all of a corporation's operations), self-imposed constraints emanating from within the business sector itself, and vigilant internal (business) and external (public) oversight. Indeed, the public should continue to become more intimately involved in corporate decisions that affect them, an approach that is now slowly gaining ground among these so-called stakeholders. Actively involved stakeholders will certainly promote more widespread corporate adoption of, and adherence to, environmentally sustainable practices.

However, maintaining for the long-term whatever gains have or will be made toward achieving environmental security can best be assured by the continuing widespread support of the public at large. I must stress that one of the crucial means of creating such support on a continuing basis will be through universal environmental education at all school levels — primary, secondary, and tertiary. Certainly not to be overlooked in this regard is a particular emphasis on environmental education in the business schools, where it is for the most still largely neglected — often being dismissed as 'socially responsible' business and thus only at the periphery of 'real' business. And, of course, environmental ethics is a subject worthy of expounding upon now and again from the pulpit. NGOs useful to pastors in that regard would be the World Wide Fund for Nature (Gland, Switzerland; 1961—) and the Alliance of Religions and Conservation (Manchester, UK; 1986—),

More broadly, given the reciprocal dependence upon each other of social security and environmental security, our long-term goals must certainly include: (a) alleviation of global wealth inequities; (b) promotion within the wealthy industrialized states of greater frugality and more sparing exploitation of the world's natural resources and sink capacities; (c) elimination of the many existing "undemocratic, despotic, and corrupt regimes; (d) encouragement of nonviolent means of conflict resolution; (e) bringing into balance the globally available renewable natural resources with the needs and desires of the global human population; (f) acceptance of the intrinsic rights of all living things; (g) support of multilateral approaches to addressing security issues, regional where possible, global where necessary; and (h) vigorous promotion of the rule of law everywhere, both domestically and internationally.

And finally, terrorist activities can certainly have an impact on national security. Such criminal and often nihilistic activities are most difficult to counter. The response, as I see it, must be heightened civilian police action within affected states, together with greatly improved international cooperation among the world's civilian police forces. The needed international collaboration would best be accomplished by the expansion or further establishment of such intergovernmental agencies as the European Police Office (Europol) (The Hague, Switzerland; 1992—) or a much improved International Criminal Police Organization (Interpol) (Lyon, France; 1956—). Anyone apprehended as an alleged terrorist must without question be treated humanely and afforded due process of the law. And any necessary actions of a military sort should certainly be carried out only under the authority of the United Nations Security Council.

Acknowledgments

A long journey commenced from attending conferences and seminars abroad and in India on environmental security, peace, development and economic rethinking, imbibing new ideas, approaches and views from scholars I met, and from books I rummaged through, getting articles, and papers published in various journals of international repute and writing books, to fleshing out the essence of all these I learnt into a book, continues to trailblaze a new path. Without the help and cooperation from others it could not have been possible to bring out this book. My unbounded thanks and gratitude go to Dr. Arthur H. Westing, Dr.Thomas F Homer Dixon of Toronto University, Dr. Richard A Matthew of California University who have played a key role in shaping my views and vision about the multidimensional aspects of environmental security from different perspectives. From the core of my heart I always convey my sense of gratitification to these erudite scholars for the materials, articles and books they have provided me without exhibiting even a slightest evidence of their being hesitant. My joys know no bounds when I feel to extend my sense of heartfelt gratitude to Marilyn Weiner and Hal of Screen scope of Washington for air dashing to India from such a far off place to take an interview with me for their documenting film on environmental security. It is really they who introduced me to Prof. Richard A Matthew. I am equally grateful to a host of critics of environmental security such as Daniel Deudney, Richard A Moss and Barry Buzan for having opened my horizon and my outlook to the neglected and not yet scrutinized aspects of environment, through their papers, articles, books and comments. I will fail in my duty if I do not give my heart felt and heart rendering thanks to the authorities of European University Center for Peace Studies, (EPU), Stadtschlaining, Austria and the Austrian Government for having privileged me with a much coveted scholarship to go to Austria and Hungary and to have higher studies on International Relation. This helped me a lot come in contact with a number of great scholars like Westing, Gunther Bachler and Johan Galtung whom I consider as yet the very prominent authorities in this field to have provided me different perspectives to my work. I also thank the authorities of Jawaharlal Nehru University Library, Indian Council of World Affairs, Teenmurti Library, United Nations Library and American Centre for having allowed me to work and collect materials in the past. My words are inadequate to express my humble gratitude and indebtedness to my teachers Dr. Benudhar Pradhan, Director, Gandhian Institute for Peace Research and Action ; and Dr. Asha Hans,Professor in Political Science,Utkal University, Vanivihar for their good wishes, inspiration, kind help and choicest blessings kept in store for me .

I always remain thankful to my intimate and scholar friend Kailash, for having been lavish in showering on me encomium for my little academic achievement which I feel I am not entitled to . Over and above, I always keep my ardent critics in high esteem, for it is they who through their stridency of attack, inspire me to ascend to the higher altitudes of learning.

I am extremely grateful to my parents, my life partner Reena in our eternal journey, and my two loving beautiful flowers Tutun and Situn who have always spurred me inspiringly to the ethos of trailblazing and emerging shining, and withstand difficulties in the preparation of the book. Behind all these and the sole doer, but for whom it was not possible to write such a book are Paramhansa Prajnanananda and my Gurudev Paramhansa Hariharananda. I deem it a God given priviledge to sing in paeans of their Beatific Bliss and Glory.

Tvameva vidya dravinam tvameva
Tvameva sarbam mama deva deva.

Thou art my Mother and Thou art my Father,
Thou art my Companion and Thou art my Friend,
Thou art my Knowledge, Thou art my Wealth,
Thou art everything for me, my Lord of all Lords.

Narottam Gaan

1

Environment from an Evaded Periphery to a Focused Primacy in World Politics

Introduction

In the halcyon days of the cold war, the concept of national security was fashioned by the West in response to a broad spectrum challenge from the former Soviet Union ranging from military to ideological, social and economic. Two mutually exclusive and anti theoretically opposed system of political economy over the future of industrial society enmeshing the two superpowers in a scuffle resulted in scripting the cold war. What was at the inception a wide conceptualization of security boiled down to a military construct under the pressure of a nuclear arms race marked by rapid, sustained, and strategically important development in technology. The Western countries under the leadership of the United States were insistent that the rest of the world follow the western liberal political economy, the espousal and implementation of which will redound to western people's extravagant and inordinate life style, consumption, and production of wealth. Reliance on military power and prowess was strategically considered important to ensure supply of raw materials from the South, and create there markets for their products, by erecting impregnably walls of tariffs, other rules and regulations against the products of the South. On the other hand, the Soviet Union's was a state interventionist and state controlled political economy finding in the politically intrastate strifes and economically poverty stricken regions of South the groundswell of revolution. Its strategy was 'export of revolution' to the South with the backing of unrivalled military strength. Where as promotion of an inordinate consumption, commodious and luxurious living, and life style and production of wealth constituted the contour and contents of western and American national security, the Soviet Union's was negatively speaking, the spread of revolution to other parts. In the other words, the entire universe, its elements, components and raw materials are to be domesticised by the west for its own interests by the spread of its liberal political economy and its associated philosophy and principles. In contrast, the Soviet Union tried to universalize its domestic political party manipulated revolution as the only reality to be theatered everywhere.

Right through to the 1980s, the arms race and the theories of deterrence interwoven with it dominated the security discourse of both the superpowers. One wanted to domesticise the universe, the other to universalize its domestic working class militancy. In the middle of the eighties, a new awareness emerged to questioning the relevance of mindless military build up as a means to define national security on a maximum –minimum continuum paradigm. In the United States a discontentment of the public brewed over the military expenditures that could be expended on maximizing their domestic imperatives in terms of inordinate consumption, exorbitant life style and production of wealth. Similarly in Soviet Union, questioning the logic of military expenditure without providing the minimum needs to its people came to the fore with the winding down of its bamboo or iron curtains.

Besides, the growing awareness that war was disappearing from the scene and in some cases had disappeared, as an option in relations amongst a substantial group of states, caused a decline in politico military construct of national security issues. The emergence of détente and effectiveness of nuclear deterrence between East and West took kindly to the outlook that the Soviet Union could be included in the emergent security core group consisting of Western Europe, Japan, and North America an outlook that grew stronger with Gorbachev enthroned in power in Soviet Union supporting an explicit demilitarization of the cold war. With Vietnam debacle driving a nail in the coffin of American supposed military arrogance and supremacy, a growing tendency in the west to question whether war was a cost effective method for achieving what said above a wide range of political and economic objectives, took strong roots. In the broadening of this outlook war as a predominant apparatus of national security faded away to question the realistic assumption about the primacy of this military apparatus in the national security construct.

Territorial exclusivity, boundary making and spatial distancing were thought no longer the key words dominating the realist assumption of security. The security shifted away from boundary or territory to people, individuals, the economy and the interspaced and external environment. What had traditionally been thought of as low politics : the international economy and the environment rose to the core of the securitization. Only during the last decade of the cold war when it was to be flickered out, the wider agenda reemerged to give prominence to the non-military aspects of national security.

In case of the environment the concept 'securitization' can be harked back to 1960s when a number of popular books such as Rachel Carson's controversial 1962 best seller 'Silent Spring' riveted to the growing impact of humankind on environment calling for a rethinking of the entire established relationship between nature and civilizations. What was a background constant in terms of a bountiful heap of lifeless raw materials to be transformed endlessly into consumptive parts of human commodious living, became a foreground variable to redefine and broaden the national security. The hitherto conventionally understood and misperceived relationship between man and nature is not just mechanical but more complex and delicate. Viewed from a Judeo-Christian perspective, and putting nature at the beck and call of human rapacious behaviour for mere exploitation starkly put future of humanity at stake.

By 1970, the observance of Earth Day, was the largest environment demonstration in history because it could spur millions of Americans to participate in this event smacking of their rising awakening to the impact of industrialization on human life, the psychological stresses of nuclearisation, environmental disasters, advancement in scientific knowledge and the compatibility of environmentalism with other anti war and women's movements.[1] This movement could point to the issue of environmental politics finding their way into the national security agenda. Further, a host of books by writers such as Paul Ehrlich (1968), Garret Hardin (1968), Barry Commoner (1971), Donella Meadows et al (1972), and Lester Brown (1972) erupted a public debate about the issue such as exponential population growth, the tragedy of commons, the negative externalities of production technology, the potential limits to industrial growth and the complex global interdependencies of the late twentieth century. These authors could be able to adorn the environmental politics with new analytical and normative discourse. In recognition of the growing importance of environmental issue forming the core of international debates, the United Nations Conference on the Environment held in 1972 in Stockholm put the environmental issue at the core of the agenda of world politics. Even during the century prior to 1972, many governmental and nongovernmental member of international community met sporadically all ineffectively to discuss a range of environmental issues.[2] It was at Stockholm an international and official status was accorded to environmental issues through the creation of United Nations Environment Programme (UNEP). Since then taking a leaf from this, the number of agreements, conventions and treaties among and between states on air pollution, climate change, forest, ozone layer, the use of nuclear weapons, protection of flora and fauna, transboundary flow of hazardous materials, has grown up as yet with many states failing either to sign or implement these agreements. A corpus of international environmental law now exists to guide and monitor state actions. The United Nations has acted to incorporate environmental law issue into its specialized agencies such as UNDP, World Bank, International Labour Organization, World Health Organization, International Maritime Organization, UNESCO.

Public concern about environmental issues climbed so high backed by many scientific findings about air pollution, deforestation, and acid precipitation that compelled state policy makers to take environmental issues seriously. The result was the creation of World Commission on Environment and Development in 1983 chaired by Gro Harlem Brundtland, former Prime Minister of Norway, and the release of its report 'Our Common Future'.

Ensconcing its attention on the global and interlocking process of population growth, food production, ecosystem protection, energy use, industrialization and urbanization with a wide ranging proposals and recommendations woven into its broad framework. The concept of sustainable development, it defines it as "to meet the needs of the present without compromising the ability of future generations to meet their own needs".[3] The elaboration of sustainable development bridging the diverse interests of developed and developing states, found its culmination in 1992 conference on Environment and Development held in Rio. This has also been integral to discussion at other UN conferences on issues such as population and development held in Cairo in 1994. Thus from Stockholm to Rio's Agenda 21, the world has traversed a long way during these last decades to include the impending environmental issues into its security agenda in recognition of the gravity and magnitude of the crisis.

In sum, it can be said that what was a moral outrage at the impact of pesticides on human health grew steadily during these years into a wide range interconnected issues, including climate change, biodiversity, resource depletion, pollution and the threat from meteorites. The underlying problem was a combination of burgeoning human numbers and rising industrialization within a finite planetary ecosystem. The entire global concern for environmental issues can be split into two : one was the Arcadian desire to securitize the environment and to make a restitution of what has been done to the environment. The second was a pragmatic survival thinking that if human beings exceeded the carrying capacity of the ecosystem in too many ways, they would peril the supporting conditions of their own prosperity, civilization, and possibly existence. Aware that the despoliation of nature in combination with growing population could strike a threatening blow against humanity making it vulnerable to disruption in production, and economic sufficiency, the entire political segment of the world decided to give primacy to the environment – no longer the evaded sector. In other words, three broad and interconnected issues emerged to occupy the attention and concern of practioners and policy makers of world politics : environmental ethies, sustainable development and environmental security.

Given the severity of the crisis, what it needs is a systemic change or an overhaul of the system that is solely responsible for such an environmental impasse. The way it has entered into the academic realm gaining legitimacy and into the mainstream of international relation shows that policymakers and scholars are very much keen on providing compromised, and short-term solutions to maintain the status quo based on a symptom cure paradigm. That means they are interested in removal of symptoms not in cure of the roots. As alleged by some critics like Vandana Shiva, Maria Mies and Carolyn Merchant including the author, this symptom cure approach has suited the interests of the rich industrialized North. By co-opting environmental issues into their hegemonic mainstream of international relations, as dependable and subordinate variables, these rich industrialized countries have successded in marginalizing the concern of the poor developing world and at the same time in exaggerating its contribution to the environmental crisis. In this way, this policy of cooption and shifting of the responsibility of the America led North countries put a veneer on their own historic nuisance to the environment. These approaches, further divested the environmental movement of their potential revolutionary ethos which are essentially required to herald systemic changes.

In the other extreme is a minority of thinkers like Julian Simon, Herman Kahn, Ronald Bailey and Wallace Kaufman who challenge the mainstream thinking of environmental politics. In their views, the assertion that certain human activities affect the environment in adverse ways that threaten both the existence and welfare of human beings, and the ecosystem's complex evolutionary process, is ill founded and exaggeratingly alarmist. These critics rely on scientific uncertainty and inconclusive studies and findings of the academic world to contest the grounds on which an overarching primacy has been put on environment over and above other factors.

In the middle of these expostulations is the environment-conflict linkage approach made by Homer Dixon and others. From an array of case studies such as in Rwanda, Somalia, Pakistan, Philippines they unearthed the environmental causes of conflict. According to their studies, environmental scarcity of resources has produced many social effects in terms of economic decline, poverty, decrease in agricultural productivity, migration and weakening of state structure which in turn result in violent intra or inter state conflicts such as simple scarcity, group-identity conflict and insurgencies and relative deprivation conflicts. These studies are diagnostic in the sense that these have brought to the center stage the environmental causes of conflict which are independent variables not subordinate to what conventionally held as primary - political, social and economic factors. Central to their model was the notion of an ingenuity gap or a disparity between the solutions required to cope with environmental scarcity and the human, social and institutional capital that could be mustered to provide these solutions. This ingenuity gap undermining human adaptive responses in combination with resource scarcity generates violent conflicts in a number of regions. Critics of this model like Richard A. Matthew, Ted Gaulin and Bryan McDonald disagreeing with what environment- conflict linage models propound, concede that "environmental stress only becomes an important explanatory variable when the level of environmental stress exceeds the adaptive capacity of a state". In their view, "this rarely happens because societies almost always find ways to cope".[4] These scholars also view that human adaptive capacity is neither completely stalled out as suggested by Homer Dixon[5] nor is it ubiquitous as suggested by Julian Simon and Herman Kahn.[6] "Rather in the vast majority of cases, human adaptive capacity lies somewhere between two unrealistic extremes".[7]

Even if adaptation means withstanding the repercussions of environmental scarcity or nuclear weapons by suffering or starving then, it does not in any way diminish the gravity of threats posed by either the nuclear weapons or the environmental scarcity of resources. In another sense, if adaptation means co-opting by finding alternative means, or changing life styles, then environmental scarcity comes as a new security thinking challenging the established mode of living, or pattern of development. Fearing the use of nuclear weapons or the devastating consequences of war on its people, a state may change its behaviour pattern and course of action in response. This perceived changed behaviour in the state does not denude the nuclear weapons of their threatening capability. Similarly, if environmental scarcity and degradation of resources envisage changed behavioural pattern, mode of living and alternative patterns of development, which is termed as adaptation, then the causative role of environmental scarcity in heralding such changes stands valid and reinforces its threats posture like the military or nuclear weapon remaining always a threat despite its being not exercised. Hence the author in this book accepting the stand taken by Homer Dixon enlarges its scope by conceptualizing that environmental security addresses the very roots of what wrecks such disastrous consequences on humanity. Neither Dixon, nor Matthew has touched the ethnocentrism of Western pattern of development, its life style and approach to life and nature. Endorsing the stand taken by Dixon and others the author holds that environmental scarcity of resources can cause intra and inter state violent conflict in interaction with its various social effects. But violence is not a solution to the crisis. In order to ensure positive peace, environmental security envisages finding out alternatives, and overhauling the dominant pattern of development and life style that has been responsible for environmental degradation. If adoption means finding out alternatives, changing the established mode of thinking and behaviour and taking steps to remove the causes of environmental scarcity of resources, then this also comes within the scope of what is defined as environmental security.

Disagreements over causes, effects, and responses as to the environment exist with differential focus prioritizing their environmental concerns. Deep ecologists advocate a biocentrism and holistic approach in which all forms of life are intrinsically valuable and interrelated. It is preservationist and concerned with defense of wilderness, ecosystems and biospecies. It is perceived as a social being solely concerned with the health of plants and animals. It puts blame on technology, pride and greed which are the forces encouraging mankind to exploit, alter and try to control the great web of life of which it is a part. A product of nineteenth century preservationist battles, deep ecology failed to cast the ecological crisis within a wider social context. Its ecological holism describes, humanness as a mere biological component of the interdependent biosphere has no necessary purpose but to serve the cycle of life; as they are at the top of the food chain, humans do not appear very important to the long-term sustainability of life. In other words "deep ecologists can be accused of dissolving humanness, either through biological reductionism or through the doctrine of larger self".[8] Indian philosophy expresses, in the heightened stage of God realization, the individualized self is transcended into universal or larger self in which the realized one perceives both the living and non living not hierarchically but with an equanimity of vision as the manifestation of God. Here the universal self is not an extension of Hobbes or Morgenthau's man. It is a transcendental or enlightened one encompassing within it the entire universe and in the universe his/her self magnified. The doctrine of larger self does not mean only that it sprewns on attribute of sacredness around the non human living and non living, but it elevates human beings to a higher plane of feeling the universal self so that he/she could perceive the animate and inanimate as nothing but his/her own self. Environmental security as defined by the authors stresses on this point to have an ethical resurgence in human beings.

Social ecologists on the other hand adopt an anthropocentric perspective. They understand and address environmental change in terms of its relationship to forms of social injustice and violence. They entwine the fate of the people with the environmental degradation. Social ecology and its ecoanarchist branch under Murray Bookchin unlike the deep ecologists stressing on unicellular earth and its ecological holism, put emphasis on political and economic aspects of the ecological crisis and construction of an ecologically proper ethic/polis theory for human kind.[9] Inspired by Peter Kropotkin's rejection of the facile interpretation of Darwinism as a theory of struggle in nature, and his advocacy of self empowered commune, Bookchin argued for ecology of freedom which is a defense of both the individual and community. The liberty of one is ensured by egalitarian and democratic relationship with the other. Solution to the ecological crisis is quite dependent on the democratic reconstruction of society, with a sensibility of sociality to freedom, peace, and ecology extending to a municipal conferderalism which will bind the sovereign communities through non representational democratic rules. The respect for life, human and non-human is premised on his advocacy of equality of unequals and recognition of differences not indiscriminate attribution of rights in the name of justice.[10]

Ecofeminism, a branch of social ecology drawing impetus from the coincidence between women's subjugation and the control of the nature, in both cases by men, aims at disabusing the political theorists of a world view of both women and nature as useful and expendable matter to be controlled, exploited and discarded to satisfy the male needs. It defends a particular constituency by reflecting on the modern killing of nature, on the deadly cumulative impact of science, monopoly capitalism and the global states system".[11]

Different Perspectives on Environmental Scarcity of Resources

The study of the relationship between population growth and environmental scarcity resources goes back to the days of Confucius and Plato. Neo-Malthusians which include biologists or ecologist argue that the natural resources are finite and put limits on the growth of human population and consumption. In their analysis, poverty and social break down result from these limits being exceeded. Taking the clue from the core idea of Malthus, the economists opine that population growth causes a drop in per capita food production. If the supply of cropland is static, and with population growing, at some point the extra increment of food population by an additional person – that is, the marginal product of labour – start to fall. This means the food production falls below the amount of food the additional person eats.[12] According to some economists, the burgeoning population can cause poverty by diverting capital from savings and investment to consumption. This diversion dwindles the long-term productivity of the economy.[13] Inducing changes in agrarian structure, the population growth constrains food production. Other also point to the role of the larger population in causing environmental degradation such as soil erosion, and in turn, poverty. Empirical studies do not stand in support of this neo-Malthusian arguments. Rather, technological change and greater inputs of capital have dramatically increased lobour productivity in agriculture ... change in agrarian food output; and larger population can lower environmental degradation."[14]

Economic optimists explaining better the adaptability of the growing population to resources maintain that capital growth arises from increased savings, and higher productivity from technological innovation. As studied by Jacob Schmookler, profit opportunities are the causal stimulus for technological innovation.[15] These opportunities are often sharpest in imperfectly competitive markets where entrepreneurs can make profit from monopoly control of new technologies".[16]

The induced innovation theory developed by J.R. Hicks in the 1930s[17] and later taken up by Yujiro Hayami and Vernon Ruttan[18] and applied to the development of agriculture in the United States, Japan and other countries, shows that changes in factor endowments of land, labour and energy are reflected in market price signals. These signals stimulate technological innovation and loosens constraints on growth. They argue that demand for new institutions is largely determined by changes in factor endowments.[19]

The economic optimists' view has three important implications. The first is that natural resources are more homogenous than commonly thought, since the right institutions and technologies allow substitutions among resources.[20] As a result, Harold Barnett and Chandler Morse note the "reservation of particular resources for later use may contribute little to the welfare of future generations".[21] Human ingenuity will eventually discover a boundless source of energy. In continuation of their arguments, they further opine that "advances in fundamental science have made it possible to take advantage of the uniformality of energy/matter—a uniformity that makes it feasible, without preassignable limit, to escape the quantitive constraints imposed by the character of the earth's crust".[22]

The second implication is that human species is biologically exceptional and it is creativity and adaptability that distinguish human beings from other animals. Archaeological evidences suggest that land degradation and over population helped tumble down many civilizations such as Sumerian, Mayan and Easter Island Civilizations. But these economic optimists argue that the limiting factors in these cases were knowledge and institutions, not resources.[23]

Third, is their assumption that resource degradation and scarcity are not problems of excessive growth of either population or consumption, but of the failures of government policy and markets. In the case of the failure of the governments to set up economic mechanism to incorporate the social costs of scarcity, resources will be overexploited and creative solution to scarcity undersupplied.[24] The analysis of Ester Boserup that land scarcities induced by population growth can give farmers incentive to raise crop yields, is found to be weak.[25] Further Alan Olmstead and Paul Rhode find that 'long run relative land prices were generally moving in a direction counter to what Hayami and Ruttan claim and in the wrong direction to explain mechanization in the period before 1910". While concluding they state that " the evolving structure of America agriculture cannot be explained simply in terms of the relative supplies and prices of a few factors", but some factors like settlement patterns, education levels and social structures are to be considered as causal variables.[26] Walking in the footsteps of economic positivists, distributionists emphasize how social and institutional arrangements within the society not the availability of natural resources are the key to development. James Boyce's study of factors influencing food production and poverty in Bangladesh represents distributionist view.[27] He discounted Malthusian explanation of poverty in Bangladesh. In his analysis population growth has actually induced agricultural innovation during the 1970s and 1980s which in turn fostered an increase in agricultural yields. He pointed to the binding technological constraint on agricultural productivity. He further argues that country's distributional inequalities have stood in the way of the process of social and technical innovation which Bangladesh so urgently needed in the 1970s and 1980s. "In the absence of a broad political mobilization of the landless and near landless majority in the countryside", such institutional changes are unlikely 'and the barriers to this are formidable'.[28] For explanation of the relationship between resource scarcity, population growth and prosperity, the social and political factors are bound to be dependent on, as pointed out by Boyce.

From the above analysis it is seen that it is not the scarcity of resources but the technological innovations, social change and human ingenuity and adaptability are very important key variables to explain the linkage between resources scarcity, population growth and prosperity. The changes that are taking place so rapidly to the ecology and planetary system, are not so simple and straightforward as to be subjected to technological innovation, or social and political manipulation, or scientific niceties, as pointed out by neo Malthusian economists, economic positivists and distributionists. Ensconcing excessively faith in human ingenuity and technological innovations has not yet halted the downward slide in environmental degradation with the dangers of climate change, ozone hole and global warming looming large on the entire living world. What remains dominant in this mainstream economic thinking is that resources are either finite or infinite depending on the way human ingenuity or technological innovation turns the finite into infinite by finding out substitutes. There is no consideration of any value or ethic that can be attached to nature except that it is to be infinitesimally exploited for the consumption of human beings.

Ecological economics

A new perspective in the form of ecological economics has emerged taking sustainability as its main nuance to challenge many of the fundamental aspects of neoclassical economic theory. The central aim of ecological economics involves depicting the relationship between "dynamic economic system and larger dynamics, but normally slower changing ecosystem".[29] In its normative sense, this relationship describes socioecological context wherein " (1) human life can continue indefinitely, (2) individuals can flourish, and (3) human cultures can develop; but in which human activities remain within bounds, so as not to destroy the diversity interests and function of ecological life support system".[30] In this definition is implicit intergenerational equity, meeting basic needs and integrating concern for ecological process into economic calculations.

The notion of intergenerational equity challenges some of the central assumption of neo-classical economic thinking The main efforts of ecological economists are directed at understanding how to approach growth policy that would not impoverish the future.[31] Another broad thrust in ecological economics is the argument that operationalizing sustainability demands redirecting economic and environmental policies to ensure a non-depleting stock of natural capital.[32]

"Environmental valuation has emerged as the primary method of accomplishing such a transition by inserting awareness over ecological entities (species, ecosystems) and services (air and water quality) into the calculus of economic well being".[33] Costanza advocates a biophysical basis for value that argues societal values should be based on the degree to which a particular evolutionary product (resource) is organized in terms of availability for human use.[34] Monetary valuation of nature as propounded by ecological economists entails the danger of reducing complex ecological process and richly realized meaning to the crude urges of money.[35]

Science : Its understanding of Nature, Resources and Human Societies

The above three camps neo-Malthusians, economic positivists and distributionists except the ecological economists tend to view that the availability of resources are fully subject to determination by human beings, institutions, policies and technologies. But the scientists' findings during these years show the other way that the influence of the complex interactive and interdependent systems of environmental resources on human societies is in part a function of the systems' intrinsic character. The threshold, interdependence and interaction of the environmental sectors such as water, forest, climate and oceans have an independent and negative impact on the availability of resources.[36] The earlier scientific view that Earth's environmental system is quite resilient, stable and prone to equilibrium in the face of human attacks on it, seem to have shifted with the present stand that the response of these environmental systems to human intervention is slow, and incremental for a long time" and then jump into drastic changes in their character on non-linear lines with feedbacks not on linearity, disjointed or separatedness functionality which are typical of the neo-classical economists, and western liberal market economy and their world view. The geographer William Clark discusses these effects in interlinked physical, ecological and social system :

Typically in such systems, slow variation in one property can continue for long periods without noticeable impact on the rest of the systems eventually, however, the system reaches a state in which its buffering capacity or resilience has been so reduced that additional small changes in the same property, or otherwise insignificant external shocks, push the system across a threshold and precipitate a rapid transition to a new system state or equilibrium.[37]

Focusing his study on the operation of the North Atlantic conveyor, Wallace Broecker reaches the conclusion that human induced global warming can cause disruption to the conveyor with potential effect on Earth's entire climate.[38] The Antarctic ozone hole stands startlingly a witness to how environmental systems can respond non-linearly to human actions. In the 1980s a linear relationship between chlorofluorocarbon (CFC) emissions and ozone depletion was accounted for what has happened to the ozone layer. What was then not remotely anticipated and has now come to surface by the scientists that when certain temperature and light conditions interact with particular concentration of water, nitrogen compound and CFCS, destruction to ozone layer can occur at lightening speed.[39]

On the basis of the new findings of the scientists about the interactive relationship between environmental systems and human society, the ecologists now hold that "metaphors of stability, equilibrium and balance are not appropriate to describe complex, interdependent systems of organisms. Instead, metaphors of anarchy, flux and constant turmoil are more apt".[40]

According to Homer Dixon, three such important factors are particularly important: extreme events, the ramifying character of scarcities in interdependent environmental system, and interactions among human impacts on these systems. He stated :[41] First, Assuming that the salient features of the environmental systems change very slowly and linearly, the frequency of events within the system can increase sharply. For example, a slow two or three degree increase in warming of global temperature might not seem too significant in itself, but for agricultural production it would prove devastating in terms of drought, floods, heat waves and storms. T.M.Wigley points out that severe social impacts of environmental change can stem "not so much from slow fluctuations in the mean, but from tails of the distribution, from extreme events". He further calculates that "a change in the mean by one standard deviation would transform the 1 in 20 year extreme to something that could be expected perhaps 1 year in 4, while the 1 in 100 year extreme becomes a 1 in 11 year event."[42]

Second, the environmental systems starting from climate to forests, soil, water and fisheries are dynamic and interdependent. Over exploitation of one resource can lead to damaging effects on the system. The diverse role of forests in providing wood for fuel, construction and paper, in reducing the amplitude of variation of the hydrological cycle by slowing runoff of rain water, and by absorbing and recycling some of it through transpiration, in stabilizing soils and reducing erosion, in absorbing and fixing atmospheric carbon dioxide and in providing a habitat for diverse organisms, are well known and beyond any dispute. If forest loss is widespread, its ramifying effects on all can be disastrous. Robert Gottfriend writes : increased use of one ecosystem good or service (function) often affects the supplies of other ecosystem functions. The relationship between these functions can be modelled in terms of key variables related to ecosystem management."[43]

Third, multiple human impacts on the dynamic and interdependent character of environmental systems can interact to produce synergistic outcome. An agricultural region may, for example, be simultaneously stressed by degraded soil and changes in precipitation caused by regional deforestation or climate change. The total effect on crop output of these interacting impacts can be much greater than the sum of their separate effects.[44] As a study shows, as an evidence of interaction effects, the combination of a warming climate with acid rain has increased the penetration of ultraviolet-B (UV-B) radiation into the lake waters of northwestern Ontario. While concluding the scholars in this study stated that much of the damage done to aquatic life in these lakes that was originally attributed solely to acidification, might actually be attributable at least in part, to increased exposure to UV-B.[45]

As the above analysis shows, neither the economic optimists, nor distributionists have taken full account of the recent findings made by scientists about the interaction, interdependence and threshold of environmental systems working not always on linear lines. Many neo-

Malthusians have just dovetailed these finding into their general alarmist notion about the human destruction of the environment.

Extrapolating insights from the galaxy of celebrated scholars in this field starting from Arthur H Westing to Homer Dixon, Myers, Ullman, Matthew Arnold, Deudney and others, I have tried to provide a theoretical scaffolding to what is usually defined as environmental security, in the second chapter of the book. While relying on arguments adduced by Arthur H Westing, Homer Dixon, Libiszewski and others, I agree with them that environmental scarcity of resources along with social effects can cause violent intra and interstate conflicts. The consequences of environmental scarcity of resource on human society are more devastating than those of war. Attribution of national security status to environmental degradation and scarcity of resources is the point of disagreement between what I propound in this book and Deudney and other critic's outlook and views. My arguments are never confined to the dominant environment and conflict linkage paradigm. This book tries to cure the root causes of the diseases and not the symptoms. My project of a theoretical build up of environmental security traverses beyond what Homer Dixon and Deudney or Matthew Arnold suggest in their studies.

In the third chapter I have made a review and rebuttal of what Deudney and other realists, neorealists and liberals have argued against the idea of bestowal of national security status on environmental degradation .The fourth chapter delineates the roots causes of environmental degradation in various sectors. The fifth chapter deals with various social effects in terms of decline in agricultural productivity, economic decline, migration and weakening of social and political institutions with reference to the findings of Homer Dixon, Boge and Libiszewski and my findings in my other works. The sixth chapter gives a portrayal of the violent environmental conflicts both intra and inter state that arises directly from environmental scarcity or degradation or indirectly in combination with other social effects; citing many case studies.

In the seventh chapter is found the arguments made by me against the western tendency to define environmental security as an appendage to or an extension of its mainstream geopolitical economic thinking, from a South's perspective. Environmental security as propounded by me is not a Western metaphor or euphemism to reinforce its hegemonic political and economic thinking, its pattern of development and exorbitant life style. The sinister move to clean the environment in the South for assurance of supply of raw materials to the North is not what comes within the ambit of environmental security.

The challenges of climate change due to emission of CO_2 to the atmosphere- mostly a doing of the rich industrialized North – to the sovereign state system, its legitimacy and the security system built around it are dealt with in the eighth chapter. In this book are found the environmental challenges calling for a review of Westphalian sovereign state system.

What is conventionally ignored, sidetracked and not taken account of, in the works of the proponents of environmental security is the challenges of environmental degradation to ethnocentrism of western pattern of development, the profligate life style and consumption of the rich industrialized countries, their approaches to life, world view and the view of the relationship between man and nature. The ninth chapter deals with all these giving an envision of a normative reconstruction of political and economic thinking, nature of man and the established relationship between man and nature on the anvil of ecological insights and spiritual essence of religions. Redefining national interests in congruence with new planetary ethics and wisdom derived from ecological insights, pragmatic view of the world and spiritual essence, and rebuilding individuals not on Hobbesian and Newtonian lines as atomistic, self moving and self directing particles of motion, but on a foundation of a spiritually informed expanded or larger self, find a detailed and critical mention in this chapter. This chapter distinguishes extraordinarily the aim, and objective of this book from those of the other proponents of environmental security and addresses the policy makers and the world leaders to eschew their traditionally held views about what constitutes national security, national interests, and development. To save the Earth and ensure the survival of human kind, what is needed an ethical resurgence in both the individuals and world leaders. This normative approach clashing with the realist paradigm seems to show the becon light to all the members of the planet trying to find out an escape route from the environmental labyrinth in which they are today. The referent objects of security are no longer the state, its boundary, but individuals, people, humanity, air, fisheries, land

and the future generations. Environmental security standing on the confluence of sustainable development and environmental ethics pleads for a resuscitation of an enlightened, broader and enlarged self. In the conclusion it can be said that a mere depiction of the environmental conflict linkages in various parts of the globe or finding out the environmental causes of conflict will not provide solution to the conflict. 'Traversing beyond' is the path that is shown in this book to find a resolution of these conflicts by removing their causes. This is not only negative, but positive in the sense that it aims at a strategy of reconstruction or deconstruction. What remains conventionally dominant or hagemonic in our political and economic thinking and world view needs to be deconstructed and a reconstruction of our political and economic society needs to be founded on the coping stone provided by this book, so that positive peace and security can usher in the world.

The unidimensional view of human being forming the core of the dominant western and modern world view of political and economic thinking needs to be replaced by a view which holds human being not a bundle of appetites, and body and mind informed desires, but an expanded universal self in which he/she realizes, the entire ánimate and inanimate world is nothing but his/her self. Since human being is the star actor, and a dominant, intelligent, and conscientious player, to move beyond the limits and trammels of the established view of human beings is possible only when the third eye opens up. Reality is not what the naked eyes see according to positivism and behavioural science, but what the closed eyes see with the vision of the third eye is reality. This is addressed to in this book. Without a normal and ethical reconstruction of human beings, the efforts to save the planet will result in fiasco. The immortal words of ancient Indian sages and saints 'Basudheiva Kutumbakam' (The whole world is like a family) resonate invigoratingly in the pages of the book with new emphasis and imperatives to replace the western preached globalization-an euphemism to universalize its political economy as the only road to Mecca of development .My work is not a deification of the mainstream economic political paradigm pontificating the entire world to act according to its dictates, but a proselytization in terms of finding alternative and diverse pathways to define development-security and environment nexus. "Do not go where the path may lead", wrote Ralph Waldo Emerson, "Go instead where there is no path and leave a trail". All civilizations, all cultures and societies in the world today need to be trailblazers in carving out different ways and paths.

All said above does not mean that this book resolves all the debates that have rankled the minds of scholars, researchers and social scientists about what role the scenarios of environmental degradation play in our world view and, political and economic thinking. Divagating from the traditional ethos of the dominant worldview, this book tries to stalk on a different path not on a trodden furrow.

References

1. John Mc Cormick, Reclaiming Paradise : The Global Environmental Movement (Bloomington : Indiana University Press, 1989), p. 47.

2. Lynton Keith Caldwell, International Environmental Policy : Emergence and Dimensions, 2nd edition (Durham, N.C. : Duke University Press, 1990) , pp. 30-54.

3. World Commission on Environment and Development, Our Common Future (Oxford : Oxford University Press, 1987), p. 8.

4. Richard A. Matthew, Ted Gaulin, and Bryan Mc Donald, "The Elusive Quest : Linking Environmental Change and Conflict", revised manuscript submitted to Canadian Journal of Pol. Sc : December 21, 2001, File no. 01-12, p.10.

5. Thomas F. Homer Dixon, The Ingenuity Gap, (New York : Knopf, 2000).

6. Julian Simon and Herman Kahn, The Resourceful Earth : A Response to Global 2000, (New York : B. Blackwell, 1984).

7. Richard A Matthew, et al, "The Elusive Quest", op.cit, p. 14.

8. Eric Laferriere, " Emancipating International Relation Theory : An Ecological Perspective", Millennium Vol.25, no.1, 1996,p.58.

9. Murray Bookchin , The Philosophy of Social Ecology : Essays on Dialectical Naturalism (Montreal and New York, NY: Black Rose Books, 1995).

10. Bookchin, The Ecology of Freedom : The Emergence and Dissolution of Hierarchy, Revised edition (Montreal and New York, NY: Black Rose Books, 1991).

11. Vandana Shiva, Staying Alive : Women , Ecology and Development in India, (New Delhi : Kali for Women, 1988).

12. Harold Barnett and Chandler Morse, Scarcity and Growth : The Economics of Natural Resource Availability (Baltimore : Johns Hopkins Press for Resources for the Future, 1963).

13. A.J.Coale and E.M. Hoover, Population Growth and Economic Development in Low-income Countries : A Case Study of India's Prospects (Princeton, N.J : Princeton University Press, 1959).

14. Thomas Homer Dixon, Environment, Scarcity, and Violence (Princeton, New Jersey : Princeton University Press, 1999), p. 30.

15. Jacob Schmookler, Invention and Economic Growth (Cambridge, Mass.: Harvard University Press, 1966), p. 199.

16. Gene Grossman and Elhanan Helpman, Innovation and Growth in the Global Economy (Cambridge, Mass. : MIT Press, 1991), p. 335.

17. J.R. Hicks, The Theory of Wages (London : Macmillan, 1932).

18. Yujiro Hayami and Vernon Ruttan, Agricultural Development : An International Perspective (Baltimore : Johns Hopkins University Press, 1971, 1985).

19. Vernon Ruttan and Yujiro Hayami, "Toward a Theory of Induced Institutional Innovation", Journal of Development Studies vol.20, no.4, 1984, pp. 203-23.

20. H. E. Goeller and A.M.Weinberg, "The Age of Substitutability", Science 191, no.4228, 20 February 1976, pp .683-89.

21. Barnett and Morse, *op.cit.*, p.10.

22. *Ibid.*, p.11.

23. Daniel Hillel, Out of the Earth ; Civilization and Life of the Soil (New York: Free Press,1991); Clive Pointing, A Green History of the World : The Environment and the Collapse of Great Civilizations (New York : Penguin, 1991).

24. Theodore Panayotous, "Roundtable Discussion : Is Economic Growth Sustainable ?" Proceedings of the World Bank Annual Conference on Development Economics 1991 (Washington, D.C. : World Bank 1992), p. 357.

25. Uma Lele and Stevene Stone, Population Pressure, the Environment and Agricultural Intensification : Variations on the Boserup Hypothesis, MADIA Discussion Paper 4, Washington, D.C.: World Bank,1989).

26. Alan Olmstead and Paul Rhode, "Induced Innovation in American Agriculture : A Reconsideration", Journal of Political Economy 101, no.1, 1993, pp.105, 110.

27. James Boyce, Agrarian Impasse in Bengal : Institutional Constraints to Technological Change (Oxford : Oxford University Press,1987).

28. *Ibid.*, p. 255.

29. R. Costanza, H. Daly and J.Bartholomew, "Goals, agenda and policy recommendations for ecological economics" in R. Costanza ed., Ecological economics : The Science and Management of Sustainability (New York : Columbia University Press, 1991), pp. 8-9.

30. *Ibid.*

31. D. Bromley, "Searching for sustainability : the poverty of spontaneous order", Ecological Economics 24, 1998, pp. 231-32.

32. P.Ekins, "Sustainability First" in P. Ekins and M.Max Neef, eds, Real Life economics : Understanding Wealth Creation (London : Routledge, 1992), pp. 412-22.

33. Christopher S. Sneddon, " Sustainability in ecological economics, ecology and livelihoods : a review", Progress in Human Geography Vol.24 No.4, 2000, p. 527.

34. Costanza, *op.cit.*, p. 334.

35. D. Harvey, Justice , nature and the geography of difference (Cambridge , MA : Blackwell,1996), p.156.

36. Thomas F. Homer Dixon, *op.cit.*, p. 37.

37. William Clark, On the Practical Implication of the Carbon Dioxide Question (Laxenburg, Austria : International Institute of Applied System Analysis, 1985), p. 41.

38. Wallace Broecker, "Unpleasant surprises in the Green house", Nature 328, no. 6126, 9 July 1987, pp.123-26 ; See also Wallace Broecker and George Danton, "What Drives Glacial Cycles", Scientific American 262, no.1, January 1990, pp. 48-56.

39. See Schneider, "Global Warming : Are We Entering the Greenhouse Century ?, (San Francisco : Sierra Club Books, 1989), p. 226.

40. Homer Dixon, Environment, Scarcity, and Violence, *op.cit.*, p. 39 ; William Stevens, "New Eye on Nature : The Real Constant is Eternal Turmoil" New York Times, 31 July 1990, p. B5, national edition.

41. Homer Dixon, Ibid., pp. 40-41.

42. T.M.L. Wigley, "Impact of Extreme Events", Nature 316, no. 6024, 11 July 1985, pp.106-7.

43. Robert Gottfried, "The Value of a Watershed as a Series of Linked Multi Product Assets", Ecological Economics 5, no.2, 1992, pp. 145-61.

44. Robert Chen and Myron Fiering, Climate Change in the Context of Multiple Environmental Threats, Research Report RR-89-1, World Hunger Program (Providence, R.I.: Brown University, 1989).

45. D.W. Schindler et al., "Consequences of Climate Warming and Lake Acidification for UV-B Penetration in North American Boreal Lakes". Nature 379, no. 6567, 22 February 1996, pp. 705-8.

2

Environmental Security : A Theoretical Quest

The late 1970s witnessed a spurt of activity resulting in a myriad of calls to redefine security in non military terms that include rising poverty, rapid population growth, spread of infectious diseases, and environmental degradation. The realist understanding and definition of security was state centric connected with state secrecy, nuclear and military power. The realist framework of national security as a function of the successful pursuit of interstate power competition through military means continued right through the cold war. According to Dalby, "cold war versions of security have usually been understood in spatial terms as moves of exclusion and protection is a spatial exercise in distancing and boundary making".[1] This means in geopolitical parlance, the spatial limitation of Soviet political control. During the last half century, security has been primarily a matter of concern for states and their military alliances. The legitimacy of the governments is generally understood in terms of provision of internal security for the inhabitants residing within its territorial jurisdiction. Where stability of the regime has been held synonymous with national security serious inroads into the realm of human rights, and internal repressions have been rationalized in the name of national security. Where political instability has been identified as threat, national security is understood as containing and limiting political change and legitimizing the status quo. The intensification of cold war climaxed in the proliferation of nuclear weapons and in the Reagan administration's Strategic Defence Initiative and Star wars merely exposed the paradox of security where the enhancement of nuclear power does not mean the enhancement of security to the people. The problems plaguing the entire humanity in terms of grinding poverty, burgeoning population growth, spread of diseases, scarcity of resources and environmental degradation could not be comprehended within the realist framework based on state centric military capability. Thus, the dominant realist paradigm to understand the current realities seems to be inadequate. It has "become an anachronism that has lost much of its explanatory and prescriptive power."[2]

Redefining Security

Understanding the increasing irrelevance of realist conceptions of security to the changing international scenario a number of approaches in the eighties and nineties redefining security in a broder and comprehensive term emerged to catch the attention of analysts.

Perspectives from South

If security meant during cold war period, the spatial limitation of Soviet influence, then what was secured is an inordinate energy and resource conserving industrial capitalist model of development and western political system[3]. From the perspectives of the South, the conventional definition of security was seen as synonymous with the preservations of US hegemony, the security of the west and the interest of the international capitalism. Going beyond the realist thinking, security was defined in terms of threats to established values and identities. Adding an economic dimension to security, some defined it not only in terms of internal security but in terms of secure systems of food, health and trade. That means national security was redefined in terms of a basic human needs provisions, denial of which to the people, will create unrest in the society posing a threat to the legitimacy of the political system.

The traditional understanding of security in terms of geopolitical interests of the North has taken for granted that the industrial North states have a right to access to the resources and raw materials available in the South whatever claims to sovereignty governments may try to make, and that military interventions, to ensure that supplies flow are justified. In its cold war manifestations security referred to the maintenance of the political order of the western world, or more broadly, modernity dominated by the rich and powerful Northern states. It means the right to maintain Northern life style and their social and political system, that underlies them that has been the object of geopolitically formulated national security policies of Northern states[4]. While defining security the Northern states have responded always to the external upheavals as these will cause disruption in resource supply or cause migration of refugees without paying attention to the internal changes that are required of them in their own societies. The persistence of focus on security in terms of geopolitical interests concentrating attention on external matters rather than internal changes as the response to changing international situation, has been the fulcrum on which the definition of security in conventional terms of Northern states rests. From the perspectives of South, the security as understood by the North needs to be redefined not in terms of its geopolitical interests but in terms of the entire humanity and more in terms of the interests of the poor individuals living in South. To provide justice, and equity both intra and inter generational and in terms of resources access, distribution and development, from a Southern perspective what should remain in the centre of the redefinition of security is the internal changes required in the Northern states to inordinate consumption and life style and its pattern of development.

North's version of security

There are divergent trends in the contemporary analysis of security in the North. One is associated with common or comprehensive security. According to this trend, the state centric military notion of security is highly flawed. It adopts a multi dimensional definition of security with emphasis on security, interdependence and multiple rather than zero sum notion of security, characteristic of the conventional cold war wisdom. This line of thinking was reflected in the advocacy of common security in the Palme Commission Report titled "Independent Commission on Disarmament and Security Issues". Focusing on the vulnerabilities of contemporary societies in the face of ever advancing technologies warfare, the concept of common security enjoined on the states to abandon their unitary military measures. The conventional definition of security had become incapable of addressing the issues arising out of the complex nature of global dangers.

During the 1980s, the common or comprehensive security approach formed an important input in the formulation of security policies in the North. A 1980 Report on Comprehensive National Security to the Prime Minister of Japan reformulated security to encompass economic vulnerabilities, natural disasters and ecological imbalances.[5] In the mid 1980s, Mikhail Gorbachev's proposal for a comprehensive system of international security talked about disarmament as well as economic and ecological security.

In Europe, the search for alternatives to military confrontations led some politically active intellectuals to develop the ideas of non-provocative defence or non-offensive defence (NOD) in the 1980s.They argued for the dismantling of ideological and military blocks. Out of this movement grew the political programme of realignment.[6]

Feminists are of the view that contemporary discourses on security have neglected women making them feel insecure. Arguing that military and national security functions of the state have always been viewed from a masculine perspective, they favoured defining security in multidimensional terms which include freedom from both physical and structural violence. Exposing relations of gender inequality and including women's experiences in security analysis can help to construct the more comprehensive definition of security that many contemporary theorists are searching for.[8] There has also been a feeling that security should be reformulated in terms of ensuring protection against events that threaten to degrade the quality of the life of people. These threats include the inability to meet basic needs, environmental degradation and natural disasters.[9] Scholars have also emphasized the need to include in the definition of security economic, environmental and demographic issues.[10] All these attempts at reformulating security underline the fact that the state centric security and its military apparatus can no longer prove adequate. A multi dimensional approach to security defined in terms of military, economic and environmental concerns and their interdependence that forms the bedrock of common or comprehensive security thinking.

Neo-realist Perception

In recent years attempts have been made to re-examine security from a broader perspective taking into account the issues raised by the proponents of common security. According to Buzan, for example, the sovereign state is likely to continue to play an important role in providing security. The evolution towards strong states, most typical of the West, will result in a greater degree of security of individuals. In the context of the international system, he does not believe that the elimination of anarchy is the answer to the security dilemma. As the system moves towards mature anarchy, a more stable form of international anarchy, which is evolving along with the progress towards stronger states, would enhance international security. The assumption that strong states would provide security to individuals however has been questioned by critics.[11] When the military dimension gets priority, the economic interests of the marginalized groups like women and minorities may be compromised. The assumption that the economic security of states would be strengthened by greater interdependence with the world economy is also open to question. Many feel that this interdependence would adversely affect the peripheral states. Moreover, environmental security may be incompatible with the continued economic development of the states and of the world economy, which is required for the progress towards a mature anarchy.[12]

New World Order Approach

Many scholars envisage a new world order where security of the individuals gets precedence over the security of the state. Klare and Thomas,[13] for example, define security from this perspective which includes measures to enhance the long term health and welfare of the human family reducing human suffering to the minimum. Richard Falk identifies the new threats which cannot be combated by the state alone and thereby complicate the realist assumption of self-help. The rise of democratic transnational forces challenging the realist worldview would constitute the corner stone of global security. This shift from national security to global security would be the core of the world order perspective. This is a human vision of security which starts with the individual.

Emancipatory Approach

There is a strong feeling that individuals, not states are referents of security. While rejecting the realist paradigm that views security in terms of division, Booth, for instance, envisions security from a holistic perspective which would integrate the best in the realist approach with a politics of emancipation that presupposes a democratic form of human security.[14]

Human Security

In his Millennium Report, United Nations Secretary General Kofi Annan noted that while security policy had traditionally focused on the defence of territory from external attack, it has now come to embrace "the protection of communities and individuals from internal violence".[15] This focus-on the protection of individuals rather than borders-is the central element of what has become known as "human security". Ideally, policies that seek to strengthen national security should also enhance human security, and vice versa. But while protecting the territorial and political integrity of states from external attack may be a necessary condition for the security of the citizens of those states, it is not a sufficient condition. So, in practice, human security's focus on the security of individuals has also meant focusing on a quite different source of security threats than those that have absorbed generations of international relations and traditional security scholars. Today, the major threat to human security derives intrastate not interstate wars. The new academic research on intra-state war focuses on the security-development governance nexus. Prescriptively, while not eschewing forceful responses to violence, it stresses non-coercive prevention policies-from preventive diplomacy, conflict management and resolution strategies to security system reform and the promotion of good governance, sustainable and equitable development, and democratization. While all proponents of human security agree that the referent object of security should be people rather than states, there is little agreement over the nature of threats that should be included. The broad concept of human security encompasses a much wider range of threats to persons than does the narrower and more traditional concept. These threats include hunger, disease and pollution and involve the denial of the most basic human needs. Indeed, in its broadest formulations, human security becomes equivalent to human well being.[16]

Environmental Security

As gleaned from the above approaches, the state-centric and military perspective of security has lost both practical relevance and intellectual credibility in the context of a number of historic forces and events in the contemporary age. These include the end of cold war, global integration of national economies, erosion of national identifies and cultures, the shift in priority from military rivalry to economic competition, and the diminishing role of state as the dominant actor in international politics. On the other hand, there have emerged diverse new issues-ranging from poverty to refugees crises, information privacy to cyber-terrorism, environmental problems to natural disasters– which requires non-state and non-military policies and strategies. These newly emerging security concerns have been characterized as non-traditional, and are now considered a major component of what is christened as comprehensive security. Among these emergent problems replacing the threat of East-West ideological divide, military aggression and struggle for global preponderance is the global environmental crisis. It looms large in terms of global warming, sea level rise, acid rain, greenhouse effect, diminishing capacity of the agricultural system, depletion of earth's finite resources, punching holes in the ozone layer, and biodiversity loss. Simply put, the global agenda has expanded since the demise of cold war as has the need for urgent attention to these problems for solution. It is thus seen that "welfare not warfare, will shape the rules and global threats like ozone holes and pollution will dictate the agenda."[17] It is within this context that the environmental question has gained worldwide significance as a security issue. In fact, environmental security stands out as perhaps the most widely debated issue, especially due to its all-pervasive nature, cross national scope and inter generational implications.[18]

While many of the past, present and future causes of conflict and war may seem to have little or no direct connection with the environment or with resources, a strong argument can be made for linking certain resource and environmental problems with the prospects for political frictions and tensions, or even war and peace. At the centre of ongoing debate is the assertion that resource scarcity and certain forms of environmental degradation are important factors contributing to political instability or violent conflict at local, regional and interstate levels. There is a growing perception that local, regional and global environmental deficiencies or resource scarcities will increasingly produce conditions that may lead to conflict.[19]

This conceptual shift towards environmental security in the context of collapse of Soviet Union and diminution of security threats associated with cold war, pitchforked into the mainstream security debates urging the policymakers to cast about for a new security focus. Despite contentious debates within the field of environmental security, a wide variety of recent comments by senior diplomats and policy makers are symptomatic of the recognition that issues related to environmental security have ascended to the highest levels.

In 1987, the General Secretary of the Soviet Union, Mikhail Gorbachev stated :

[The World] is not secure in the direct meaning of the word when currents of poison flow along river channels, when poison rains pour down from the sky, when an atmosphere polluted with industrial and transport waste chokes cities and whole regions, when the development of atomic engineering is justified by unacceptable risks... The relationship between man and the environment has become menacing. Problems of ecological security affect all the rich and the poor.[20]

In November 1989, Prime Minister Margaret Thatcher gave a speech to the United Nations General Assembly-saying :

While the conventional political dangers – the threat of global annihilation, the fact of regional war- appear to be receding, we have all recently become aware of another insidious danger. It is as menacing in its way as these more accustomed perils with which international diplomacy has concerned itself for centuries. It is the prospect of irretrievable damage to the atmosphere, to the oceans, to earth itself.[21]

In April 1997, as United States Secretary of State, Ms Albright went on to say:

Not so long ago, many believed that the pursuit of clean air, clean water, and healthy forests was a worthy goal, but not part of our national security. Today environmental issues are part of the mainstream of American foreign Policy.[22]

It was in this backdrop of future non-military threats that a number of studies and research institutions undertook works to determine if and how environmental factors were related to the intra and inter state conflicts or violence that appeared to be surfacing around the globe.[23] These studies varied in their approach, their terminology, and their specific findings and consistent in their explanations suggested that environmental stress, operating through a set of intervening variables, could contribute to violent conflicts, and thus constituted a threat to national and international security. Furthering this argument that the mounting environmental problems and the associated issues need an attention with which the conventional outlook of national security appears to be incongruent, Jeremy Rifkin points out: "The environmental threats facing the planet are not simply the result of scientific miscalculation. Nor are they merely the consequences of ill-conceived management decisions. Ironically it is the notion of security upon which our entire morden worldview is based that has led us to the verge of ecocide... In less than a century the practice of geopolitics thus pushed the world to the brink of both nuclear Armageddon and environmental catastrophe forcing us to reconsider the basic assumption of security that animate the morden world view"[24].

Similarly, Lodgaard feels that " the concept of environmental security challenges established frames of mind and political conflicts. It conveys a message that environmental problems have a legitimate claim for status as military problems have."[25]

Another proponent of environmental security in the same vein opines:

...national security is not just about fighting forces and weaponry. It relates to watersheds, croplands, forests, genetic resources, climatic and other factors that rarely figure in the minds of military experts and political leaders, but increasingly deserve in their collectivity to rank along side military approaches as crucial to a nation's security.[26]

The growing significance of environmental security is quite evident in major international forums. Some well known examples include the UN Conference on Human Environment (1972), Ottawa Conference on Conservation and Development (1986), United Nations Conference on Environment and Development (1992), World Summit for Social Development (1995) and UN Symposium on the Global Environment in the 21st Century (1997). These conferences and symposia eventually resulted in various international conventions and protocols for environmental protection or security, such as the Vienna Convection for the Protection of the Ozone Layer, the Convention on Biological Diversity, the Framework Convention on Climate Change, the convention to combat desertification, and the Kyoto protocol.[27] Harking back to as far as 1987,for example, The World Commission on Environmental Development (Brundtland Report) entitled 'Our common Future', stresses the factor of environmental degradation causing the violent relationship between states:

"Environmental stress is both a cause and effect of political tension and military conflict...nations have fought to assist or resist control over raw materials, energy supplies, land, river basin, sea passages and other key environmental resources. According to the report, such conflicts are likely to increase as these resources became scarcer and competition for them will increase."[28]

Environmental security as a concept encompassing non-military aspects was officially mentioned for the first time in the International conference on the Relationship between Disarmament and Development convened by the UN General Assembly in New York from 24 August to 11 September 1987. The final document adopted by consensus states:

Recently non-military threats to security have moved to the forefront of global concern. Underdevelopment and declining prospects for development as well as mismanagement and waste of resources, constitute challenges to security. The degradation of the environment presents a threat to sustainable development.... Mass poverty, illiteracy, diseases, squalor and malnutrition affecting a large proportion of world's population often become the cause of social strain, tension and strife.[29]

Though a number of studies have spawned up recently demonstrating the linkage between environmental stress and conflict, there has been some criticism of these findings as happens in other fields. Critics like Deudney argued that the language of security— a discourse marked by militarism, competition and secretiveness is incompatible with the ethos of modern environmentalism.[30] With the end of cold war and collapse of Soviet Union, the exposure of secretive nuclear installation to international scrutiny, and security being associated not with state, but with the individuals or people, the language of security is no more counched in terms of militarism, competition and secretiveness, rather in terms of demilitarization, nuclear free world and public democratic governance. The dominant geopolitical paradigm of national interest was based on realist parlance of US and its allies' interests overwhelming the entire globe since the second world war. When environmental security because of environmental catastrophic consequences threatening the existence of human beings leapfrogs to the center stage of national security concern, it as assumed, is not to be based on same dominant geopolitical paradigm of North's interests to acquire the status of national security. Rather it calls for a redefining the very foundation of traditional outlook of national security based on geopolitical interests of the dominant players in the international realm. If it is to be just an appendage to the geopolitical interests of the dominant players defining the world order since the second world war, then environmental security is not in any way different from conventional security based on secrecy, militarisation and competition for dominance. It has climbed up to the top of the national security agenda not on the basis of realist paradigm, but on a different paradigm. That what cannot be described in old traditional languages lacks the raison detre to be called national security stands discredited in the face of new kinds of threats that are stemming from unprecedented environmental crisis. Advocacy of hesitancy to term environmental security as national security means advising to navigate the uncharted seas of environmental crisis with a warped, discredited and conventional realist cartography. If in the words of old ideas of security, environmental security is tried to be defined, the very dilution of what it stands for, takes place, and the perpetuation of the geopolitical interests of the dominant North led by US continues to the perils of the people in the developing South. When the discourse of environmental security is defined in terms of old conventional idioms i.e. the unhindered right of dominant players to resource supply sources in the South, it merely 'extends the explicitly geopolitical understanding of security that were widespread during the cold war to maintain the current global division of wealth."[31]

Another objection is that in the concept of environmental security the absence of "us vs them" which is a cardinal part of traditional concept of national security, is like, according to Deudney, taking sex out of 'rocks and roll'. In his words, 'rock and roll' was originally coined as an euphemism for sex.[32] Interstate or intrastate conflicts over environmental scarcity of renewable resources like water, and fish is a clear-cut example of 'us vs. them' syndrome. There is no dearth of evidences to show that water sharing has been a major cause of conflicts between and among nations.[33] For example, despite a set record of tension and on being war path between Lebanon and Israel, recently they were on loggerheads with each other when Lebanon began a project to increase water supply from the Wazzani to its neighbouring village. Israel is opposed to the project as a major share of water for the nation's main water reservoir, the lake Tiberias comes from the river. Reacting to the project, the Prime Minister of Israel, Ariel

Sharon said : it will create grounds for war between the two nations."[34] The very presence of Bangladeshi refugees in Indian North-eastern states and Assam due to environmental scarcity of land in Bangladesh, has been an armed conflict involving the military forces of both India and Bangladesh.[35]

Two distinct features of environmental security are : First, the environmental causes of conflict, i.e. environmental factors behind potentially violent conflicts; Second, the impact of environmental degradation on overall political economy, health and life of the people. If environmental degradation or deficiencies create the conditions that render conflicts, act as multipliers that aggravate core causes of conflict[36] or act as a catalyst factor in creating conflicts, then environmental degradation or scarcity of resources becomes a national security issue.[37]

There are some scholars who refuse to accept the links between environmental degradation and national security threats[38]. They argue that "taking the term security and using it to encompass anything that threatens our well-being, might really take the term and dilute it of any significance"[39]. The feeling is that defining environmental issues in terms of security threats may contribute more to the militarisation of the environment than to the greening of the environment."[40] In other words, defining environmental issues in terms of security risk is in itself a risky operation ... We may end up contributing more to the militarisation of the environmental politics than to the demilitarization of security politics".[41] This kind of view is harboured only when environmental security is understood by the North as to include its unimpeded right to access to resources of the South. That means, environmental security is considered as mere an appendage to the geopolitical security interests of the North defined in terms of realist paradigm making a great departure from what the proponents of environmental security originally understand it, and shifting very conveniently the onus of environmental protection to the South, so that the flow of resources from the South remains unhindered. Thus, the relationship between resources and development should also form part of the security debate. From the cold war perspective, security issues were understood in terms of the right of access of the developed world to the resources of the South. The major problem in the current environmental crisis is the absence of equal access to resources.[42] The world economic structure and major international financial institutions function in a way as to serve the economic interest of the developed North reinforcing the widening North-South economic disparities. They also facilitate the continued exploitation of the environment and natural resources of the South. With the perpetuation of this kind of unjust economy coupled with debt crisis, the environment will be seriously affected rendering those belonging to the poor nations more insecure.

The cold war mindset premised security largely on the perpetuation of existing flows of resources from the South to the North. Such thinking was predominant in American policy-making, particularly during the second Gulf war, which was waged to protect American interests. Behind the military onslaught of America against Iraq at present is heard the loud concern of US for the treasure chest of Iraq's oil on which Saddam Hussain sits. Conflicts or wars over access to resources are in line with the traditional paradigm of security based on the state's military supremacy. Hence, the need to rethink the concept of development. For, the Western paradigm of development based on unrestrained industrialism, fossil fuel technology and extraction of nonrenewable resources have polluted and degraded the renewable resources like water, land, air, forest and biodiversity in terms of greenhouse gases, global warming, acid rain, desertification and deforestation. As a result, the environmental scarcity or degradation of resources, has emerged as a threat to the security of a large number of people. The conflicts that may arise out of such types of threats are different from the traditional conflicts over access to resources. Thus, the environmental security assumes the character of national security not in terms of traditional justification for more military supremacy and state power, but in terms of the need for an overhaul of what developed North has so far defined as development to resolve the environmental crisis.

There is almost an unanimity among the scientists about their predictions of threats of environmental degradation to the security of the people. The worst affected will be the poor people in the South. But scientific inaccuracy, lack of certainty about the impact of environmental degradation region wise and lack of precision, associated with the environment are cited to argue against the environmental security. Neither the actual increased exposure to UV rays from the punching of the hole in the Ozone layer nor the degree of harm it will wreak on plants, animals and humans can be calculated to a mathematical accuracy. Neither the eventual increase in global average temperatures from a given level of greenhouse gas emissions nor the impacts on weather patterns, human health, crops or sea level rise on region or countrywise has been studied with a scientific precision. Similarly, the rate of land degradation on a global scale cannot be measured with accuracy and its impact on future food productivity can be predicted with any scientific precision. Thus, each of these environmental threats to global well being has been subject to significant empirical and scientific uncertainty and inaccuracy. All said does not anyway denude environmental security of its significance. Such uncertainties highlighted to refute the basis of the concept environmental security are also found in case of military planning based on worst-case contingencies which are most unlikely to occur. If the very fear of possible threats from the other states' use of weapons conventional or nuclear could be the basis of the realist framework of national security even without having the remotest chance of exercising military options, why the real dangers of environmental degradation cannot be held justifiably as threats to the national security of the people! Yet military preparations are justified for such future apprehended threats from other countries as an insurance against these unlikely threats.[43] With threats of environmental degradations to the people looming large, the concept 'environmental security' stands the logic to replace the traditional realist framework of national security defined in terms of military insurance against future uncertain and unlikely threats.

There is a feeling among scholars that a rethinking of the concept of security keeping in mind environmental consideration could become a dangerous tool in the hands of the totalitarian left who could attempt environmental collectivism.[44] It is also argued that since the logic of ecology has religious dimensions, the handling of environmental issue carries the danger of eco fascism.[45] The danger of environmental collectivism or eco fascism, however, seems exaggerated. For, a resolution of the ecological crisis requires a rethinking of the entire development strategy. Addressing the security concerns of the people taking into consideration all their dimensions rather than of the state alone, runs counter to the very rationale of relating security to the state military power. Where there has already been growing demands for more democratization, people's participation, diversification of state power, human rights and global governance and more transparency in decision making, the logic of

Eco fascison or totalitarian left sounds little credible. Since these regimes regard themselves as the arbiter of national security, they may oppose many international environmental movements, rules and regulation, on the pretext that these constitute sources of threat to national security.[46] But in the face of the deepening socio-economic crisis due to environmental scarcity or degradation of resources, these regimes may find the going tough. Since ecogeographical regions do not always coincide with territorial units, on many issues the priority cannot be given to strengthening state power but to environmental co-operation among states.[47]

It is therefore not suprising that a strong feeling is growing that environmental security is a national security concern. Although many scholars think that environmental security is part of the economic agenda, the concept of national security has been tacked on to environmental debates, because it has a mobilization potential.[48]

Indeed, if the predictions of scientists about climate changes in terms of global warming and greenhouse effects come true, then fears regarding the break-up of some Antarctic ice sheets could put "environmental security at the top of the global agenda very soon".[49]

Environmental Conflict

In spite of these criticisms, the most influential arguments in the field have concluded that environmental stress is linked to conflict indirectly, but significantly. While the precise character of the linkage varies from case to case, it is made possible environmental stress is supposed to overwhelm the adaptive capacity of some societies while simultaneously introducing and reinforcing forms of conflict and instability. While the field of environmental security is characterized by division and debate, there is a strong degree of theoretical consensus among researchers. The most prominent and dominant thread that runs through the entire environmental security literature is that the environmental stress simultaneously reduces adaptive capacity and engenders conflict.

The dominant paradigm in the environmental security literature harks back to writers like Thomas Malthus, but it has obtained its contemporarily influential formulation in the works of Thomas Homer Dixon and Toronto Group. From numerous of empirical case studies conducted by them on

Haiti, Chiapas, Gaza, Pakistan and Bangladesh – India, they concluded that environmental scarcity of renewable resources viz., water, forest, fisheries and cropland give rise to a number of deleterious social effects including economic decline, social segmentation and human migration – these social effects in interaction with other political, economic and social factors, can generate conflict and instability. Central to their model was the notion of supply of ingenuity gap or a disparity between the solution required to cope with environmental scarcity and the human, social and institutional capital that could be mustered to provide these solutions. This ingenuity gap essentially undermines human adaptive responses. That means, according to their studies, resource scarcity and limited or inadequate adaptive capacity coalesce together to generate violent conflicts in various regions.[50]

Dominant Paradigm

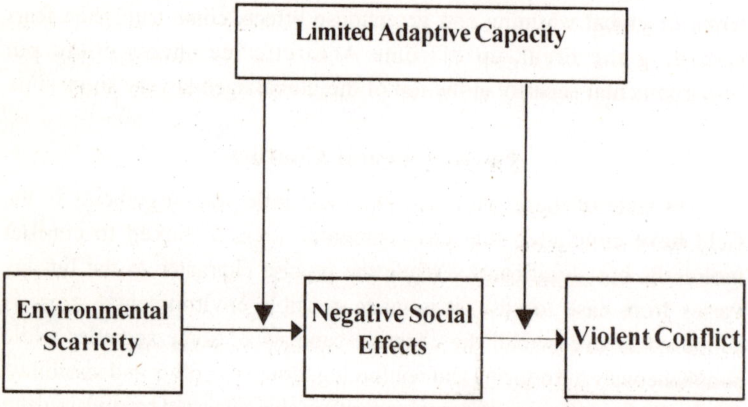

A theoretical representation is made in this figure[51] to understand the linkage between environmental stress and conflict. On theoretical ground it is criticized that even if resource scarcity plays a causal role in conflict, there are other more convincing explanations for conflict. Further, the prospects for adaptation, cooperation or resolution are not as completely bleak as is often presented. At the very prospects of an unlikely nuclear war, the fear of destructibility by the use of nuclear weapons has not any way diminished the universality of nuclear weapon as a main national security issue. It is that fear of destruction which may have brought many nuclear powers to negotiating table for cooperation or disarmament.

The resultant cooperation should not be cited to underrate the holistic perception of nuclear weapons as being a national security concern. Similarly, environmental scarcity of resources with its disastrous social effects may not be everywhere a direct cause of violent conflict, sometimes in combination with other factors causes conflict. For example, sea level rise a consequence of global warming due to North's mindless pursuit of industrialization based on fossil fuel technology will affect low-lying areas such as Bangladesh below the sea level turning millions of them as refugees fleeing to Indian territories that is likely to exacerbate the already infested ethnic conflict in North-Eastern states. That it would not affect other countries does not any way diminish the environmental causes of conflict not less than the universality of nuclear weapons as a security concern. The consequences of environmental scarcity of resources on political, economic and social life of a state or states may prompt states or nations to cooperate to avoid the negative effects of war on environment. So the environmental co-operation syndrome as argued by some should not be overrated to diminish the role of environmental scarcity of resources causing conflict or violence. No nuclear war has ever been fought in this world after American's dropping of hydrogen bombs on Japan's Herosima and Nagasaki during the course of Second World War. This would be as preposterous to say that nuclear weapon lack the potential or prospects for conflict as to say that the environmental scarcity of resources lacks the potential for conflict as it has not resulted in conflict in some areas or has resulted in cooperation. Adaptation means the involvement of all state apparatus, institutions, scientific community, even the social system and value structure including the affected to take steps that would obviate the negative effects of environmental scarcity. The poor developing countries may not have the economic and scientific wherewithal as argued by Dixon to cope with the situation. Even the developed countries will not find it a smooth sailing while responding to the environmental crisis. For example, in regard to stabilization of CO_2 in the atmosphere American steps to renege on its commitment agreed to at Kyoto were a capitulation to the petroleum and coal industrial lobby. What has threatened the world is the Western inordinate life style and pattern of development causing global warming, sea level rise and ozone layer depletion. To ensure environmental security what it requires is to make an overhaul of its pattern of life and development which is not possible with strong protestations from high-class people. So Bush, American President caving in to the pressure of the industrial class justified his rejection of Kyoto protocol on the ground that countenancing the Kyoto commitment would hamper its economy and life style of its people.[52] Environmental security means overhauling the entire economic development pattern and existing social values related to liberal capitalism which is not possible without politico-socio-economic disruption. So the potential for conflict lies as much in the environmental scarcity of resources as in the very remedies and adaptive measures – a part of fulfillment of environmental security.

Inspite of the criticisms above, the main thrust of the studies showing the linkage between environment and conflict, remains paramount : the environmental scarcity operates through a set of intervening variables, contexts or ideational factors that directly cause conflict, and that environmental social effects and stress, as such outstrips attempts to cope with the crisis.

Another study undertaken by Swiss Peace Institute's Environment and Conflicts Project (ENCOP) was in the direction of what the Toronto group suggests. Gunther Bachler belonging to this group in his study used the term transformation by which he meant "the introduction of a heuristic concept in recording those interactions between the three levels – nature, human beings and the economy in a regional context-which have led to conflicts between human communities and will ... increasingly lead to such conflicts."[53]

Degradation is used in this context as an indicator of the degree of environmental transformation. In the words of Kates et al, "the biosphere has accumulated or is on its way to accumulating such a magnitude and variety of changes that it may be said to be transformed."[54] In other words, "transformation conflicts are caused by fundamental anthropogenic changes in the environmental media and by the interactions between the processes of change and their consequences in the eco-regional context."[55]

Another authority Stephan Libiszewski belonging to ENCOP, making a distinction between traditional resource wars and environmental causes of conflict is of the view that "an environmental conflict is a conflict caused by the environmental scarcity of a resource, that means : caused by a human made disturbance of its normal regeneration rate. Environmental scarcity can result from the overuse of a renewable resource or from the overstrain of the ecosystem's sink capacity, that is pollution. Both can reach the stage of destruction of space of living. Conflicts caused by physical, geopolitical or socio-economic resource scarcity are not environmental conflicts but traditional conflicts of resource distribution."[56]

As found in the Toronto Group's research the ENCOP also walks on the same furrow that scarcity gives rise to conflict. Also it is alleged to be silent about human adaptation to resource scarcity. In these studies, it is mal- development, a condition similar to Homer Dixon's ingenuity gap that undermines human responses and enhances conflicts.[57] Though the ENCOP model is more complex than the Toronto group's general model, but the theoretical argument presented by both is very similar.[58]

In a 1999 NATO pilot study this dominant environmental conflict paradigm was found in the line with the other studies.[59] According to this study, "it is not environmental stress in isolation that characterizes the nature of the conflict between groups but other factors".[60] Here too political, economic and social factors intervene between environmental stress and conflict. The State Failure Task Force (SFTF), another research group marking the same dominant paradigm as with the preceding studies, reiterates the causal weight of environmental stress through intervening variables.[61] While the researches did find that massive environmental damage in a short time frame could directly cause political collapse, the far more common scenario was that environmental stress operates through the intervening variable of quality of life to generate conflict. Though highly critical of Toronto group, recent work by Gleditsch and de Soysa appears to have been co-opted by the dominant paradigm. According to their studies, resource scarcity, one component of a multifaceted development problem that they call "poverty" reduces agricultural production which in turn leads to conflict. That is, scarcity leads to a negative social outcome which becomes the proximate cause of conflict.[62] Beyond this, however, the Gleditsch and de Soyasa model is similar to the Toronto Group in that there is little room for human adaptation to conditions of scarcity. Indeed, their key independent variable, poverty, is defined by a lack of social and human capital. "Like Homer Dixon's ingenuity gap, Gleditsch and de Soyasa's explanatory variable appears to lock in negative social outcomes, leaving little room for human adaptation."[63]

There is some variation in these models and term presented by these research groups. But in each case the general theory linking resource scarcity and conflict is the same. Homer Dixon focuses on an "ingenuity gap", ENCOP on "mal-development" and Gleditsch and de Soysa on "poverty." Nevertheless, all these phrases describe a similar process- a limited capacity to respond to scarcity. The model which is critical of all the above models adopted by Richard A Matthew et al finds flaws in current theory in environmental security on two counts. "First, it focuses on too short a time frame. Second, and relatedly, it underestimates human adaptive capacity." [64] They were averse to accepting the two extremes, Homer Dixon's lack of adaptive capacity in the poor countries, and Julian Simon and Herman Kahn's faith in the infinite capacity of individuals to adapt. To them, human adaptive capacity lies somewhere between these two extremes. Julian Simon writes, "there is no physical or economic reason why human resourcefulness and enterprise cannot forever continue to respond to impending shortages and existing problems with new expedients that, after an adjustment period, leave us better off than before the problem arose." [65]

Homer Dixon responds to this saying, "while I acknowledge the extraordinary potential of human resourcefulness and enterprise, I nonetheless argue that some societies – especially poor societies – will not be able to supply the unprecedented amounts of ingenuity they will need to solve their emerging scarcity problems."[66] This clearly demonstrates, Homer Dixon never maintains that all societies will not be able to cope with the situation. It is some poor societies because of various factors will be in dearth of required supply of ingenuity to adapt to the situation. This brings Homer Dixon closer to the position taken by Richard A. Matthew and his associates. If adaptation is a dynamic and continuous process, then there may be adaptation failure in some and adaptation success in others visited with trauma, travails and hardships. When such individual trauma, travails and ordeals while adaptating to the environmental stress becomes a collective or group phenomena in a particular political context or in the face of growing incapability of state, the potential for violent conflict exists. Taking a departure from this, Matthew group holds that from a longer time perspective even if the adaptive failures through feedback will result in adaptive success, thus avoiding the prospects for environmental conflicts. Contrary to what Neo-

Malthusians such as Paul Ehrlich, Robert Kaplan, and Thomas Homer Dixon contend," adaptive mechanisms do not simply stall or become overwhelmed while levels of environmental stress continue to escalate."[67]

To quote Richard A Matthew further : "This is not to say that all forms of adaptation are socially desirable. Adaptation can mean people of the developing world going further and further into the hinterland to gather fuel, wood and water; parents sending their children to work in factories because of crop failure; and simple belt-tightening people eating less,"[68] or as happened in Kalahandi district of Orissa people eating the dried up kernel of ripen mangoes died while adapting to food scarcity, or people in these tribal areas of Orissa are forced to sell their children as bonded labourer as a mark of adaptation to the environmental scarcity of crop land and food. These latter examples are certainly unsettling and portent of potential for violent behaviour in an organized form depending on the nature of political system prevailing. If to use the language of Matthew, these starvation, death, or going without food, or half fed situations are the genuine form of adaptation, then conflict arising out of these situations created by environmental scarcity in the form of collective or group behaviour or response can be a genuine version of adaptation with a view to develop new legitimate and representative political authority and institution and new value structure that would reinvent the political-economic thinking which so far has contributed to environmental scarcity. As Walzer argues, conflict is a normal and important aspect of social development.[69] Then the portrayal of environmental crisis as an environmental conflict paradigm from a broader perspective does not lose its significance or acceptability, as a theoretical framework as argued by Richard Matthew and his group. The conclusion that this group arrived at is not to deny the linkage between environmental scarcity and conflict but stresses on "longer term studies" that "might provide deeper understanding of the ways in which concerns about environmental stress and degradation are relevant not only to the those concerned with international security, but also those worried about human rights, social justice and sustainable development.[70]

On the one side the economic deprivation, decreased agricultural production, health hazards, poverty, starvation or half fed or even death situation due to pollution and water scarcity- part of the adaptative success as propounded by Matthew – may cause indirect conflict in association with other factors as nuanced by Volker Boge[71] or Homer Dixon.[72] The other side, is the situation of direct conflict or violence arising out of the environmental scarcity of resources - a case of adaptative failure. Hence the conflict emanates directly from the overuse and or pollution / destruction of a renewable natural resources. For example, if a downstream riparian threatens to go to war against an upstream riparian, because the latter pollutes river water so gravely that it cannot be used by the inhabitants of the downstream riparian, who are highly dependent on this water, then the environmental character of conflict is evident.[73] An array of evidences can be cited to show such threats of conflict over environmental scarcity of water all over the globe.[74] That in some cases it has led to cooperation to avoid war does not divest the environmental conflict of its theoretical appeal. Another example of direct violent conflict over environmental scarcity of resources like fish can be cited. The decline in fish catch due to environmental degradation all over the globe and in South Asian regions in particular could result in a series of violent conflict between Tamil Nadu fishermen of India, and Sri Lanka over fishing in waters of Kachchativu Island ceded by India to Sri Lanka as a gesture of friendship and good will in 1974.[75]

The two aspects of environmental scarcity induced situation in terms of adaptive failure and adaptive success can dovetail together into what is called environmental security in a broader sense. What is called as adaptive success at micro level along with adaptative failure in terms of poverty, food scarcity, health hazards or suffering due to environmental degradation at macro level, for example the impact of global warming due to western pattern of development based on fossil fuel industrialization and inordinate life style, poses challenges to redefine development and the vision of and approaches to life. What remains pivotal to defining environmental security in a broader sense that will remove the causes of environmental conflict as well the causes of environmental scarcity that inflicts on the individuals in terms of health hazards, suffering, economic deprivation, poverty or starvation as undergone by them as a part of adaptation. Thus, at the confluence of the morally and ethically required nuance of sustainable development, economic rethinking, new planetary principles and new vision of and approach to life, stands the environmental security. Thus, the very basis of environmental security challenges the established frame of mind about what so far has been defined as development, pattern of life style, and geopolitical security of the sovereign state system. It is not the national security in terms of military but the planetary security that will be the foundation of environmental security on this confluence.

References

1. S. Dalby, "American Security Discourse : The Persistence of Geopolitics," Political Geography Quarterly 9, 1990, pp. 171-88.

2. O.R.Holsti, "International Systems, Systems Change and Foreign Policy." Diplomatic History 15, 1991, p. 94.

3. P.Cooke, " Nation, State, Modernity," in R.Preet and N.Thrift, eds., New Models in Geography, Vol. I (London : Unwin, Hyman, 1980), pp. 267-91.

4. Simon Dalby, "Threats From the South? Geopolitics, Equity, and Environmental Security," in Daniel H. Deudney and Richard A.Matthew, eds, Contested Grounds: Security and Conflict in the New Environmental Politics (Albany: State University of New York Press, 1999), pp. 164-65.

5. R.W. Barnett, Beyond War : Japan's Concept of Comprehensive National Security (Washington : Pergamon Brassey's 1986).

6. M. Kaldor and R. Falk, eds. Dealignment : A New Foreign policy Perspective (Oxford: Basil Blackwell, 1987); see also, D.Smith and E.P.Thompson, eds. Prospects for a Habitable Planet (Harmondsworth : Penguin. 1987).

7. C. Enlee. Bananas, Beaches and Bases : Making Feminist Sense of International Politics (London : Pandora, 1989); V.S. Peterson, ed.. Gendered States : Feminist (re)visions of International Relations Theory (Boulder, CO : Lynne Rienner, 1992).

8. Ann Tickner, Gender in International Relations : Feminist Perspectives on Achieving Global Security (New York: Columbia University Press,1992) p. 6

9. Richard H.Ullman, "Redefining Security," International Security, Vol.8, 1983, pp. 129-53.

10. Barry Buzan, People, States and Fear: An Agenda for International Security Studies in the Post Cold War Era (London: Harvester . Wheatsheaf, 1991).

11. S.Smith, "Mature Anarchy, Strong States and Security," Arms Control, Vol.12, 1991, pp. 325-29.

12. S. Dalby, "Security, Modernity, Ecology: The Dilemmas of Post-Cold War Security Discourse," Alternatives, Vol. 17, 1992, pp. 95-134.

13. M. Klare and D. Thomas, eds., World Security : Trends and Challenges at Century's End (New York ; St. Martin's Press, 1991)

14. K. Booth, "Security and Emancipation," Review of International Studies, Vol. 17, 1991, pp. 313-26.

15. Kofi Annan, We the peoples : The Role of the United Nations in the 21st Centuary, March 2000.

16. Andrew Mack, "Human Security in the New Millenium," Work in Progress, Vol. 16, No. 3, Summer 2002, p. 5

17. J.Joffe, "Entangled for ever", in C.W. Kegley Jr. and E.R. Wittkoff, eds.,The Future of American Foreign Policy (New York : St. Martin's Press, 1992), p. 35.

18. M. Shamsul Haque, "Non-Traditional Security and the Environment in North east Asia", Work in Progress, Vol.16, No. 3, Summer 2002, p. 24.

19. Peter H. Gleick, World's Water : The Biennial Report on Fresh Water Resources (Washington D.C.: California : Island Press, 1998), pp.105-106.

20. M.Gorbachev, "Reality and guarantees for a secure world", Moscow News, No.39, 1987, p. 3287.

21. Quoted in Peter H.Gleick, World's Water, *op.cit.*, p. 106.

22. U.S. Department of State, "Environmental Diplomacy : The Environment and U.S. Foreign Policy", U.S. Department of State, Washington, D.C. : 1997.

23. Ted Gurr, Minorities at Risk: A Global View of Ethnopolitical Conflicts (Washington DC : US Institute of Peace Press, 1993).

24. J. Rifkin, Biospheric Politics : A New Consciousness for a New Century (New York ; Crown, 1991), p. 2.

25. Sverre Lodgaard, " Environmental Conflict Resolution," paper presented at the UNEP meeting on Environmental Conflict Resolution, Nairobi, 30 March 1990, p. 18.

26. Norman Myers, "The Environmental Dimension to Security Issues". The Environmentalist, Vol. 6, Winter 1986, p. 251.

27. See M.Shamsul Haque, "The Fate of Sustainable Development Under the Neoliberal Regimes in Developing Countries", International Political Science Review 20/2 (1999), pp. 199-222.

28. World Commission on Environment and Development, Our Common Future [Brundtland Report (New York / London : Oxford University Press, 1987)], p. 290

29. Quoted in Dietrich Fischer, Nonmilitary Aspects of Security: A System's Approach, United Nations Institute for Disarmament and Research (UNIDIR), (Aldershot: Dartmouth Publishing Company Ltd, 1993), p.10.

30. Daniel H. Deudney, "The Case Against Linking Environmental Degradation and National Security" , Millennium 19, 1990, pp. 461-76.; Daniel H.Deudney, " Environmental Security : A Critique", in Dan Deudney and Richard A. Matthew, eds., Contested Grounds : Security and Conflict in the New Environmental Politics (Albany : State University of New York Press, 1999).

31. Simon Dalby, "Threats from the South ?", *op.cit.*, p. 160

32. Daniel H.Deudeny, "The Case against Linking" *op.cit.*, p. 467

33. Haleh Hatami and Peter H. Gleick, "Chronology of Conflict over Water in the Legends, Myths, and History of the Ancient Middle East", Report of the Pacific Institute for Studies in Development, Environment and Security (Oakland, California, October 1992). See also Peter H. Gleick, World's Water ... op.cit., pp.105-53.

34. Down to Earth , Vol. II, No.10, October 15, 2002, p. 16.

35. N.Gaan, Environmental Degradation and Conflict : The Case of Bangladesh-India (New Delhi : South Asian Publisher, 1998), see also Thomas F. Homer Dixon, Environment, Scarcity, and Violence (Princeton / New Jersy : Princeton University Press, 1999), pp. 95-96, 141-42

36. Norman Myers "The Environmental Dimension to Security Issues," The Environmentalist Vol. 6, Winter 1986, p. 253.

37. Gareth Porter "Environmental Security as National Security Issue", Current History, May 1995, p. 218.

38. See Daniel Deudney, "The Case Against Linking Environmental Degradation and National Security", Millennium Vol.19, no. 3 Winter 1990, pp. 461-76; Mathias Finger, "The Military, the National State and Environment", The Ecologist, Vol. 21, no. 5, 1991; Richard H. Moss,"Environmental Security? The Illogic of Centralised State Responses to Environmental Threats', in Paul Painchaud, ed., Geopolitical Perspectives on Environmental Security, Cahier du GERPE, no. 92-05, Universite Laval, Quebec, 1990.

39. Daniel Deudney "Rethinking the Link between Environment and National Security Issues", Perspectives on War and Peace, 1990, p. 12.

40. Hans Bruyninckx, "Environmental Security: An Analysis of the Conceptual Problems Defining the Relationship between Environment and Security",

paper presented at Budapest on the Workshop on Environment and Security organised by the European Peace Research Association, 11-16 November 1993, pp. 1-16.

41. Lothar Brock, "Security Through Defending the Environment: An Illusion in E. Boulding", ed. New Agendas for Peace Research : Conflict and Security Reexamined (Boulder Co: Llynne Rienner, 1992), p. 98.

42. R.D. Lipschutz and J.P. Holdren, "Crossing Borders: Resource Flows, the Global Environment and International Security", Bulletin of Peace Proposals, Vol. 21, 1990. pp. 121-33.

43. Gareth Porter, *op.cit.*, p. 220.

44. Painchaud, Geopolitical Perspectives, quoted in Ole Waever, "Securitization and Desecuritization", Working paper, Vol. 5, 1993, Centre for Peace and Conflict Research, Copenhagen, p. 14.

45. Andre Gorz, Ecology and Freedom, quoted in Ole Waever, *Ibid.*, p. 14.

46. See S. Dalby, "Ecopolitical Discourses, Environmental Security and Political Geography", Progress in Human Geography, Vol. 16, no. 4, 1993, p. 512.

47. See Marc A. Levy, Robert O.Keohane, and Peter M.Haas, "Improving the Effectiveness of International Environmental Institutions", in Peter M. Hass, Robert O. Keohane and Mark A. Levy eds., Institutions for Earth: Sources of Effective International Environmental Protection (Cambridge, MA: MIT Press, 1993); Katrina Rogers, "Rivers of Discontent – Rivers of Peace: Environmental Cooperation and Integration Theory". International Studies Notes, Vol. 20, no. 2, Spring 1995, pp. 17-18.

48. Barry Buzan, "Environment as a Security Issue", pp.1 and 24 f, quoted in Ole Waever, Securitization and Desecuratization", p. 13.

49. Barry Buzan, "Rethinking Security after the Cold war", Nordic Journal of International Studies Vol. 32, no. 1, March 1997, p. 12.

50. Thomas F. Homer Dixon, Environment, Scarcity, and Violence (Princeton, New Jersy : Princeton University Press,1999); Thomas F. Homer Dixon and Jessic Blitt, Evidences : Links Among Environment, Population, and Security (New York : Rowman and Littlefield, 1998); Thomas F. Homer Dixon, " Environmental Scarcities and Violent Conflict: Evidence from Cases", International Security, Vol. 19, 1994, pp. 5-40.

51. Thomas F. Homer Dixon, Environment, Scarcity, and Violence, p. 7 and Daniel Esty et. al., "The State Failure Task Force Report: Phase II Findings",

Environmental Change and Security Project Report 5, 1999, p. 63; S.C. Lonergan, ed., Environmental Change, Adaptation and Seçurity, (Boston : Klewer,1999); David Dessler, "Review of Environmental Scarcity and Violence", Environmental Change and Security Project Report 5 (1999), pp. 100-101.

52. David L. Levy and Daniel Egan, "Capital Contests : National and Transnational Channels of Corporate Influence on the Climate Change Negotiations", Politics and Society, Vol. 26, No. 3, September 1998, pp. 337-361. See also N.Gaan, "Politics of Governance of Global Climate change : Not on Equity but on North's Interests", India Quarterly, vol. LVIII, no. 4, October-December 2001, pp. 89-112.

53. Gunther Bachler, Conflict and Cooperation in the Light of Global Human-Ecological Transformation, ENCOP, occasional paper 9, October 1993, p. 5.

54. Robert W. Kates and B.L. Turner II and William Clark, "The Great Transformation", In B.L. Turner II, and William C.Clark, Robert W. Kates, John F. Richards, Jessica T. Matthews, William B. Mayer, eds., The Earth as Transformed by Human Action : Global and Regional Changes in the Biosphere over the Past 300 years (Cambridge : Cambridge University Press with Clark University, 1990), p. 1

55. Gunther Bachler, Conflict and Cooperation, *Ibid.*, p. 16.

56. Stephan Libiszewski, What is an Environmental Conflict? ENCOP, Occasional paper no. I, July 1992, p. 6

57. Richard A. Matthew, Ted Gaulin, and Bryan McDonald, "The Elusive Quest: Linking Environmental Change and Conflict" submitted to Canadian Journal of Political Science, Revised Manuscript submitted; Dec 21, 2001, File Number: 01-12, p. 7.

58. See for details Nancy Peluso and Michael Watts, ed., Violent Environments (Ithaca: Cornell University Press, 2001), pp.15-24.

59. Committee on the Challenges of Modern Society,NATO, "Environment and Security in an International Context", NATO Report 232, 1999.

60. *Ibid.*, p. 22.

61. Daniel Esty et.al., "The State Failure Task Force Report: Phase II Findings", Environmental Change and Security Project Report 5, 1999.

62. Ted Gaulin, "To Cultivate a New Model: Where do Soyasa and Gleditsch Fall Short", Environmental Change and Security Project Report 6, 2000, pp. 104-107.

63. Richard A. Matthew, et al., "The Elusive Quest", *op.cit.*, p. 9
64. *Ibid.*, p. 25.
65. Julian Simon, The Ultimate Resources (Princeton, N.J. : Princeton University Press, 1981), p. 345 ; and Julian Simon, The Ultimate Resource 2 (Princeton, N.J. : Princeton University Press,1996)
66. Thomas F. Homer Dixon, Environment, Scarcity and Violence, p. 114.
67. Richard A. Matthew, et.al., The Elusive Quest, p. 21.
68. *Ibid.*, p. 16.
69. Michael Walzer, Just and Unjust Wars : A Moral Argument with Historical Illustrations (New York: Basic Books,1992); Michael Walzer, Spheres of Justice: A Defence of Pluralism and Equality (Oxford : M. Roberston, 1993); Michael Walzer, Thick and Thin : A Moral Argument at Home and Abroad (Notre Dame : University of Notre Dame Press, 1994); and Michael Walzer, "The Politics of Rescues", Social Research 62, 1995, pp. 53-67.
70. Richard A. Matthew, et.al, "Elusive Quest", p. 24.
71. Volker Boge. "Proposal for an Analytical framework to grasp Environmental Conflict", ENCOP, occasional paper no.I, July 1992, Swiss Peace Foundation. p. 5.
72. Homer Dixon, Environment, Scarcity and Violence, *op.cit.*
73. Volker Boge, *op.cit.*
74. Peter H.Gleick, World's Water : The Biennial Report on Fresh Water Resources (Washington D.C. : Covels, California,Island Press,1998); Haleh Hatami and Peter H.Gleick, "Chronology of Conflict over Water in the Legends, Myths and History of the Ancient Middle East", Report of the Pacific Institutes of the Studies in Development, Environment, and Security, Oakland, California, October 1992; Helga Haftendorn, "Water and International Conflict", Third World Quarterly Vol. 21, no. 1, 2000, pp. 51-68; Marq De Villiers,Water Wars, London : Weidenfeld and Nicolson, 1999); World in Transition : Ways towards Sustainable Management of Fresh Water Resources, German Advisory Council on Global Change, Annual Report 1997 (Heidelberg : Springer – Verlag Berlin. 1999)
75. N. Gaan, Environmental Scarcity of Fish and Conflict: The Case of India and Sri Lanka over Kachchativu Waters". International Studies Notes. Vol. 22, No. 3, Fall 1997. pp.17-23.

3

Politics of Environmental Security and Problems of Widening : A Critique

Although many recent approaches, such as constructism, critical geopolitics, postmodernism and historical sociology have strongly disavowed the idea of 'state security' propounded by the realists, neorealists, and the neoliberals, they have not been able to dilute the centrality of security in International Relations.

The analysts favouring the realist paradigm hold that the international system is fundamentally anarchic in the sense that political life therein is dominated by states not subject to any authority superior to their own sovereignty. Within this anarchic system, states will purse relentlessly their respective national interests in order to ensure their survival in the face of formidable forces trying to deny this. The core concept of national interest is to be defined in terms of power as the only means of attaining national prosperity and security. Within the anarchic system, states are in constant competition, which, however, does not prevent them from accommodating their respective interests through cooperation where it directly serves them in their quest for national security and prosperity.

The central assumption of Kenneth Waltz's Theory of International Politics (1979) is typically Hobbesian. He wrote : "The state among states, it is often said, conducts its affairs in the brooding shadow of violence. Because some states may at time use forces, all states must be prepared to do so–or live at the mercy of their militarily more vigorous neighbours. Among states, the state of nature is state of war."[1] Waltz conceives the international system as self help system and treats states as defensive actors, in as much as their primary, though not exclusive, intention is self preservation. Thus, while states are not necessarily self-aggrandizing, they are compulsively self help units, looking for their own security amidst they are compulsively self help units, looking for their own security amidst an anarchical milieu where the probability of threats to their survival is always real. Thus states are primarily concerned about their own security and forced to view other states as potential threats. States are basically self-help units within an anarchical system devoid of an authoritative or commanding sovereign regulator.

The end of cold war and disintegration of Soviet Union in the 1990 and 1991 led many idealists to dream about the outlines of a New World Order that would give a quietus for ever to the anarchic international system. These significant events could make a toll of the usefulness of realist paradigm especially as these laid bare some of its weaknesses and contradictions like the fact that it failed to foretell or give hints about the collapse of cold war and the tumble of mighty Soviet Union.[2] But the euphoria about the envision of a New World Order did not last long and as the subsequent events such as the unexpected failure of the Somalia intervention, the disasters in the former Yugoslavia, events in Haiti and Rwanda, plight of refugees in Zaire and American military intervention in Iraq bypassing UN, showed, the New World order, the contour and contents of which could have supposedly fleshed into a shape, instead buried in the oblivion, again giving rise to the strengthening of age old held view that the realist paradigm still dominates the international system. No doubt the delivery was a still born New World order baby but the birth pangs it left resulted in the shaking up the global agenda, the surfacing of hitherto neglected and side tracked issues such as environment, human rights, human security and protection of minorities in the center stage, resurgence of regionalism, universalization of liberal market economy and evolution of international institutions and regimes, and the growing importance of non state actors.[3] While attending to these turbulent changes in the international system, the states again struck to the pursuit of national interest as their dominant stake understood on realist paradigm. American involvement in Afghanistan crisis and in Iraq for the second time ousting Saddam Hussain in the face of stiff opposition from world public opinion, United Nations and other dominant powers such as German, France, Russia and China, only attested to this and American President Bush' rejection of Kyoto protocol reinforces the view that the realist paradigm of national interest and security remains still dominant. Over recent decades, the environmental crisis has brought to the center stage the "non military dimension of security as an important feature of state behaviourism."[4] This led to many strategic analysts to redefine the concept of national security as to broaden its scope,[5] as the traditional approach reflected too narrow a vision not only of the problems but also of the solutions available to cope with threats.[6] A broader definition instead is felt essential that would take into consideration the fact that a threat to national security exits once an action or sequence of events "threatens.... to degrade the quality of life for the inhabitants of a state or ...threatens significantly to narrow the range of policy choices available to the government of a state or private non governmental entities within the state".[7] The preceding chapter describes in details about the efforts at defining the concept of environmental security. "The emergence of the concept of environmental security became inevitable once national security became associated with ' quality of life' within a sociopolitical context marked over recent years by the introduction of environmental questions into overall national and international concerns."[8]

Environmental Security A Dependable Variable

Realists as argued by Michel Frederick are of the view that the unreliability of international institutions, mechanism and rules, regulations and law to manage the environment has recapitulated to state's primacy in providing security to its people against the environmental threats. For example, despite existence of over twenty international and regional agreements on the prevention of marine pollution, each year finds tens of millions of tons of waste and polluting substances being dumped into the sea. Over these years adduced as a priority concern for the international community, quality of the environment has remained yet elusive and not incorporated into a genuine right to a safeguarded environment. International environmental regimes such as agreements and conventions on climate, forest, biodiversity, pollution, WTO, signed over these recent years have not been able to halt the slide in environmental deteriorations let alone manage it for human safety. Emission of carbon dioxide into the atmosphere goes on increasingly with the prospects of global warming looming large. As argued by Frederick, the unreliability of the existing mechanisms, environmental regimes and institutions for managing the environment has introduced an element of uncertainty into the international system, putting primacy on state as the only security paradigm against the perils of nature and environmental degradation. He writes, "Consequently, it is through the prism of their own national security and hence their national interest, that states will analyze the impact of global environmental problems along with state behaviour apt to upgrade the situation or further harm it. Within such a context, the environmental security concept will then take all the importance that states will effectively grant it".[9] On realist parlance this means reducing the environmental factor not an independent but dependable and subordinate variable to the politico-military construct of national security.

From the realist perspective the environmental security being a dependable variable thus takes on the meaning of security of the environment involves three elements : (1) the sustainable use of renewable and non renewable resources, (2) protection of the elements-air, water, soil- so as to prevent pollution from stifling natural regeneration; and (3) the maximum reduction of hazards related to industrial activities.[10] According to Arthur Westing, its scope covers the overall problems associated with the protection and use of the environment. In his words, "environmental security has two basic sub components. The first of these – environmental protection has three parts : protection from wartime and similar vandalism; protection from medically unacceptable environmental pollution; and protection for special areas, from all permanent human intrusions. The second sub component of environmental security is the sane resource utilization, whether non-extractive or extractive depends upon exploitation, use or harvesting at all levels and employing procedures that either maintain or restore optimal resource services or stocks... exploitation of renewable resources must be carried out strictly on the principle of sustained use or sustained discard, with exploitation of non-renewable resources strictly on the principle of frugality."[11] The use of the words "sustained use or sustained discard and frugality suggests impliedly the challenges to the established notion of development within the dominant social-cultural context and the security system built around it. At the same time it calls for an ethical resurgence in human beings impinging on them the need to moderate and delimit their inordinate consumption and life style. Thus rethinking the dominant view of development and the ethical imperatives comes within the ambit of what is redefined as environmental security, as described in the preceding chapter. The realist interpretation of environmental security ignores these imperatives for reviewing the dominant ethnocentric definition of what so far has been universalized as development. The realist view of environmental security devoid of the development - security syndrome, aims at the security of the environment for the perpetuation of the dominant paradigm of development. The realist understanding of environmental security does not envisage any plan or proposal to end its self-help, divided, hierarchical and anarchic inside/ outside construction of international system. Rather these intrinsic features of international system as outlined by neo-realists and neo liberals can be better explained by the insistence of the rich industrialized North, especially the US on security of the environment in the South and their meaningful participation for an equity based management of the crisis and sharing of the burden. But this equity was ignored while exploiting the natural resources of the earth.

The idea of environmental security stems from an essentially globalist view of inter state relation. Its main focus is on security of the planet and its resources in totality that contributes to the survival of humanity and to development perpetuity.[12]

As seen in the preceding chapter Homer Dixon and others deal with the question differently : the environment is looked upon as an independent variable, where as state security which is involved, acts as a dependable variable. Environmental security thus takes on the meaning of "the environmental component of national security."[13] These studies demonstrate that the security of the environmental crisis can be a threat to state's national security depending on the gravity of the social effects it produces in terms of economic decline, poverty, social unrest, migration and threats to national and territorial integrity..The dominant environmental conflict paradigm focuses on the conflict dimension of the inter state relations or intra state situations, which will stem from environmental antagonism which be it local, regional or global in nature.

The correlation between national security and its environmental components is seen from first, the direct environmental causes of conflict, where the environmental problems such as over water, fisheries, trans border pollutions are directly the bone of conflict between nations, second, the indirect environmental factors of conflict where environmental problems - not the main insecurity factor but the accessory insecurity factor - in association with preexisting political, economic, social and military contexts exacerbate the situation adding a new dimension to them or acting as catalysts.[14] In the words of Frederick, "this is a seductive approach because it blends well with the view of international relations that attributes rational behaviour to state entities engaged in power politics."[15]

If analysis of environmental conflict is within the classical range of military confrontation between and among states, then on realist parlance, these conflicts are in no way different from the political, economic or military factors of state security. Environmental causes of conflict are simply addition and dependable variable to the other politico-military construct of conflict. Consequently as argued by the realists, since the interstate conflicts caused by environmental crisis are state centric and responded within the military apparatus of states, the environmental security is not at all different from traditional state security. As argued in the introduction chapter, to study 'environmental security' in terms of interstate conflicts is to deal with the symptoms and not with the root causes. What has provided grist to the realists not countenancing the dominant environmental conflict paradigm is the latter's circumscribing the environmental security to the scenarios of interstate conflict within state military apparatus.

Realists Redefining Environmental Security

If security denotes a situation of tranquility resulting from an absence of danger, then from a realist perspective, this tranquility situation is valued in the context of the dominant entity, the state in the international system. International system as comprehended by realists is largely an interstate affair when matters of security are raised, they always hinge on state and pertain to the ability of states, and the societies within, to maintain their independent identity and their functional integrity.[16] According to realists survival[17] is infact the key : the survival of the unit as basic political unit (sovereign state) "security refers to those cases where a threat or development is designated as incompatible with state's sovereignty which leads to a test of will and force – there by testing whether the state is a sovereign state. If it designates a problem as a security problem and does not manage to fence it off, its status will be threatened."[18] Security is thus in their eyes an ordering of prioritization of state's will exercised upon a congeries of conditions, economic, political, social and environmental in their interactions for its survival and that of international community. The environmental condition is not the determinant of state's prioritization of will, but a subordinate and dependable factor. This is what Barry Buzan points out explaining the realists' standpoint. Each condition defines a focal point within the security problematic and a way of ordering priorities.[19] This is no doubt true that all these conditions cannot be studied in isolation from one another. But the pursuance of dominant socio-politico economic order - a western synonym for development and survival and identified often with national interest- can become a threat to the ecological fabric of the society. To address these with the military components of national security has spurred the realists to refute the arguments of the dominant environmental conflict school of thought for isolating the environmental causes from others and warn the latter avoid falling into such a trap. What came in for being widely swiped at by the realists is the dominant environmental conflict paradigm's analysis of conflict in military components of the state and segregation of the environmental component from others as the causation of conflict. As will be seen in following chapters, the case studies made by the environmental conflict schools stand a testimony to the interactive linkages between environment and violent conflict. When Barry Buzan in 'People, States and Fears' divagates from the military sector to discussing security in the four other sectors, political, economic, ecological and societal – the logic clearly says that security began as military but the military definition is increasingly challenged by the new sectors ; but what was it that made the military sector conspicuous, and what is that now qualify the other to an almost equal status ? This is not really addressed squarely by Buzan but he does actually give the answer or hints at it.[20] He argues more extensively like this :

"Because the use of force can wreak major undesired changes very swiftly, military threats are traditionally accorded the highest priority in national security concerns. Military action can wreck the work of centuries in all other sectors. Difficult accomplishments in politics, art, industry, culture and all human activities can be undone by the use of force. Human achievements, in other words, can be threatened in terms other than those in which they were created, and the need to prevent such threats from being realized is a major underpinning of the state's military protection function. A defeated society is totally vulnerable to the conqueror's power which can be applied to end ranging from restructuring the government, though pillage and rape, to massacre of the population and resettlement of the land. The threat of force thus stimulates not only a powerful concern to protect the socio-political heritage of the state, but also a sense of outrage at the use of unfair forms of competition."[21]

This used to characterize the military sector. When the same can be done through economic or political means overturning the political order these will be security problems too. Thus, the basic definition of security problems is that they can undercut the political order and there by "alter the premises for all other questions"[22] –therefore they have to be addressed before all other questions. Because if they are not ; there will not be any other questions because the unit will cease to exist as a sovereign unit. Thus, the environmental security is not mere narrative diagnosis of the causative linkage between environmental degradation and violent conflict affecting the security of the state and international system, or halting at the use of military apparatus of the state for settling the environmentally induced and caused conflict. It challenges the premises of the dominant existing socio-politico economic system that identifies with the state which have been a nuisance to the environment and ultimately to the people. The defect both with the realist and environmental conflict schools is that they incline to make the state the referent for security not the individuals, people, and the alternative premises that question the established dominant western paradigm of development and security thinking.

According to the realist framework, "environmental security represents an absence of non conventional threats against the environmental substratum essential to the well-being of its population and to the maintenance of its functional integrity."[23] On the basis of this realist definition the environmental security must be understood within a state perspective. In chime with realism, within which the concept of environmental security cannot simply be given a "security of the environment" meaning without the risk of distorting the very notion of security since the latter remains the state's prerogative. The notion of security can be understood as such under all its aspects be they political, economic, social, military or environmental. This means notion of security under state perspective remains supreme and all-inclusive and the environmental security is subsumed within it contrary to what environmental security has been defined in this work. It is the other way round that the environmental security is the overall vast design that challenges the premise on which the traditional notion of security rests and goes beyond the framework of state security. That environmental security is the key and foundation of all other security has not been recognized by realist thinking. According to Taylor, under international law there is no such thing as the idea of world environmental order common to all does not exist. The world has been divided the earth on the basis of politically earmarked territorial units. The concept of sovereignty and territorial integrity in which are ingrained the notion of national security are the very foundations of international law. In this picture of the realist territorially and politically divided sovereign state systems, "there is no direct protection of the environment per se".[24] If modernity is understood as the rendering of nature into territorially and politically carved out states, then these are incongruent with the ecogeographical region. An ecogeographical region can be terrestrial or aquatic or both. It may be conceived of essentially as an ecological subsystem made up of living and non-living components of the environment that interacts to form a life support system. It is by no means fully self-contained because it has numerous feedback relationships with subsystems in the adjoining areas and beyond. Thus, it is a part of the global ecosystem. But then it has a dynamic of its own so that it may also function independently of the contiguous and more distant regions as well as of the globe as a whole.[25] The various sorts of socially and politically determined territorial regions such as ethnic, linguistic, religious ones and so forth are superimposed upon the ecogeographical region. However, lack of correspondence and congruence between the two has often in the face of environmental degradation generated situation of conflict between and among these ethnic, religious or linguistically divided societies and states. Force or war within the military apparatus of the state has never been a solution to these conflicts.[26] So long as the root causes of these conflicts mainly embedded in environmental factors, remain unaddressed and determine what constitutes security not with state as the referent, force or war will only aggravate the situation dealing not with causes but with symptoms. The very threats that stem directly from environmental degradation and indirectly in combination with other factors are lost sight of or downplayed in the analysis of realism. On the other hand, what the realists suggest is that in the core of the environmental security states need to be given preponderance. Given the nature of the existing international system as a power game and the little role that the transnational forces, public opinion, ideologies, international organizations and small developing nations play for the preservation of the power equilibrium, to give preponderance to power maximizing states, as argued by realists is not only necessary but natural. Since the environmental crisis is global in terms of climate change, global warming, thinning of ozone layer and loss of biodiversity, the proponents of environmental security advocate not for a conventional statist military approach. But the realists are of the view that the state is the only referent for security to make its own judgement as to what constitutes the threat to its security and survival. Further, the notion of security is tied with state which is in an advantageous position to translate in terms of issues. "It can be considered that a threat to state's environmental security exists once an action or sequences of events directed in a non-military perspective against its environmental substratum threatens the quality of life of its population and / or the range of policy choices available to the government in its main areas of activity."[27]

In realist notion of security is involved subjective variation as the propensity to perceive events through the prism of a security threat varies from state to state. What is considered a threat to one state may not be a threat to another. A close link is always found between security and the population's quality and style of life, social and political way of thinking. A sovereign state's prerogative to secure that identity is what comes within the purview of realism as security. For example, the underpinnings of western power maximizing state for security are based on the western life style, pattern of development and inordinate consumption. To secure that commodious living which is the identity of western populace distinguished from that of other continents, has been the main objective of national interest defined always in terms of power which will enable it to secure resources from other parts of the world even by the use of force. The traditional wars over resources, and raw materials in the past and in the present over oil can be explained by this realist logic. What is perceived as a threat to the luxurious and rich people in the North is the scenario of non-availability of resources or any kind of compromise on their way of life. Thus, America is not in a position to reduce emission of CO_2 to the atmosphere, as it believes that it will put a compromise on their way of life. So in realism's notion of security is involved the subjective and differential analysis of security based on a hierarchical, divided and power maximizing world. All these situations presuppose a state of war as well as an armed intervention by the state under attack. Where as considerations of oil played a significant part in the security thinking of America in relation to Middle East and Iraq, denial of it is perceived as a security threat, where as other nations and world public opinion and even the UN could not buy American idioms and logic to define security threats in the case of American attack on Iraq. So in realist framework of security consideration there would be no question of perceiving or analyzing even the environmental issues other than within the diagnosis of military security. In other words, the realist notion of security cannot be greenized.

On the other hand as analyzed in the subsequent chapters, environmental security does recognize the link between population's quality of life and environment or nature. Nature is not a heap of raw materials, an object and a lifeless other to be exploited for what has been defined ethnocentrically as prosperity and development. A coordinated, symbiotic and organic view of the evolutionary process of the entire nature where both animate and inanimate play a significant role to sustain livelihood is required to found the basis of what should constitute the paradigm of development and security. It is not a hierarchically divided and disjointed world based on exploitation and inequity and widening hiatus between rich and power, but a scheme of development and security where equity is established between and among the present generations but also between the present and future generations both living and nonliving. It is not the realist notion of security that view the ecosystem, resources and natural cycles not only within its territory but also in other parts of the globe to be exploited for the prosperity and security which is the identity referent for its populace. Thus, what is sustained and conducive to security is unsustainable and security threats to other people. On the other hand environmental security challenges the very basis of the realist framework of security and its logic. It does not acquiesce in statist frame of security based on duality, division between rich, and poor, nature and the universe and the present and future generation. It is universalistic, planetary and cosmic view of security that will determine not only the national security of each and every state, but also security of the people. Contrary to this logic of environmental security, the realist definition of security centers round the state's social well being and the maintenance of its functional integrity. These two criteria are not any way to be compromised in the name of environmental security. In realist thinking these two criteria provide the frame of reference to map out the notion of security on militarist and statist denominations. If states act first and foremost in line with their national self-interest defined on these above two criteria and in terms of power and security, then pollution problems, environmental scarcity and degradation of resources such as water, fish, air, etc. can be measured in terms of national securities thus threatened. This is to reiterate what many scholars on environment and conflict such as Homer Dixon and others have tried to prove culling from many case studies that environmental security and degradation of resources directly or indirectly in combination with other factors depending on the severity and magnitude of environmental crisis, social effects and adaptive failure of the people, can contribute to violent conflict both interstate and intrastate having serious implications on the stability and security of the international system. In the realist framework the criteria of social well being and preservation of functional integrity remain uppermost while perceiving the threats from environmental degradation, where as in the analysis of Homer Dixon and others environmental factor remains preponderant and determinant to challenge the established conventionally understood meaning of security. War or use of force in response to environmental threats in terms of statist and military framework does not provide the solution and ensure environmental security. Environmental security traverses beyond this recognizing the environmental threats to security as the frame of reference to provide a second look at what so far constitute the copingstone of development, security and prosperity so that security to all-both animate and inanimate and future generations-can be ensured taking all as subjectivities important players in a cosmic vision of the universe.

Deudney's critique of environmental security

According to Daniel H. Deudney[28] in three major ways the linkage between environmental degradation and military violence can be established. The first point is that the states in the pursuit of their national security to secure freedom from violence through military means spend a lot on resources which can be diverted towards conservation of environment.

There is no guarantee that the money saved from reducing military expenditures could be spent on environment. Without any cutback on military expenditure the world can afford environmental restoration. It is preposterous to assume that the states would prefer to tighten the strings of their purses in the face of deteriorating environmental degradation in various sectors such as land, water, forest and biodiversity. Popular upsurge and mass discontentment against government showing short shrifts to impending environment crisis should not be ruled out. If the world can afford expenditures on environment even without many cuts on military expenditure then why the conditions of the environment have been getting depleted day by day?

Second, war is destructive of the environment which is both unintentional and intentional. He cited examples showing how the military destruction of olive groves in Mediterranean land, led to the long lasting destruction of the land's carrying capacities,[29] the US' bombardment and use of defoliants in Vietnam caused significant environmental damage[30] and more recently during the Gulf war of 1991, Saddam Hussian's forces set the oil fields of Kuwait on fire causing massive ecological damage and resource wastages.[31] Further extensive use of nuclear weapons would have significant ominous impacts on the global environment in terms of altered weather pattern i.e. nuclear winter and depletion of the ozone layer.[32] Even the conventional war in industrialized countries can be environmentally devastative with release of radiation from civilian nuclear plants.

Third, preparation for war and production of both nuclear and conventional weapons have also deleterious effects on environment. Literatures are not wanting to show how "over the last century, the nuclear weapon states, particularly the United States and the former Soviet Union, generated enormous quantities of radioactive waste as a byproduct of the nuclear weapons development and production."[33] According to Deudney as put forth by the proponents of environmental security, the resources and costs used for preparation and use of nuclear and conventional weapons can be invested for improving the environmental conditions. The costs involved in terms of loss of life and destruction of the environmental biodiversity questions the very logic and raison detre of the international system and state system's dependence on use of force and violence both nuclear and conventional for resolution of conflict. Nobody can deny the fact that war and preparation for it has environmentally ruinous effects. Production and use of weapons of mass destruction has been a major cause of environmental degradation which in turn comes as a predominant threat to human life. Environmental security deals with all the causes of environmental degradations out of which war and preparation for it is one. The proponents of environmental security depict analytically showing many case studies how the environmental degradation leads to violent intrastate and interstate conflict. Besides, the present work while defining environmental security stresses on rethinking the socio-political basis which has given rise to such a mindset and established framework of realism looking at the rest other than the self as the other to be exploited for one's own interests. No doubt, elimination of war as a means of resolving conflict would not bring an end to environmental degradation in other sectors. What environmental security suggests is to rethink the western pattern of development, extravagant life style and consumption and approach to life that has given rise to such a mindset about war, the relation between man and man, man and nature and the present international system.

To equate environmental degradation with diseases, aging, crime or accidents as equally security threats as done by Deudney is highly erroneous. Diseases, aging, crime etc are natural processes of human evolution. Even in the absence of environmental degradation, decay, diseases and aging will occur to all forms of life whether human or non human. If these occur naturally and by the process of evolution, nobody considers these as security threats to human beings. The way Deudney applies this logic to question why environmental degradation can be considered a security threat shows that he failed to show the difference between the natural process of evolution, and the human being made disturbance into this natural process. Environmental degradation is not natural it is anthropogenic and created by human beings, their way of life, pattern of organization, development and social –political –economic thinking. Studies show, environmental degradation and scarcity of resources and their social effects have intensified and spread the frequency of diseases, created new diseases, brought aging faster than its natural occurrence, increased the number of accidents, crimes and violent conflict. Consequently, there is nothing wrong in terming these recrudescence of diseases, aging, accidents, crimes and violence as security threats if factored by environmental degradation occupying the status of national security not in the sense of being required to be resolved within the realist military apparatus of the state. If the main objective of national security is to provide a good life and establish social cohesion and stability, then these threats assume very much the character and status of national security. Environmental security is thus an assurance of human security against the environmentally induced diseases, aging, crime and violence in a broader and more comprehensive sense occupying the status of national security not through military means as these however invincible these may be, cannot stop their recurrence, hastening and recrudescence. Despite changes and growing interdependence due to revolutionary steps in the field of communication to end distance, the irony is that the international system is tried to be understood in Westphalia logic to fit into western strait jacket in the 21st century. Persistence of this attitude is found in Deudney when he opines that "there is nothing distinctively national about either the causes, the harms or the solutions that warrants privileging the national grouping" on the grounds that environmental problems, causes, and effects are confined to the boundary of a state.[34] He has ignored how the environmental degradation or development activities in one state has been a security threat to another nation. For example, the scientists predicted sea level rise submerging low-lying areas of Bangladesh, Maldives, Egypt affecting millions of people is not the doing of these affected countries. Sea level rise is the result of global warming caused by the rich North countries' mindless historic emission of CO_2 to the atmosphere in pursuance of their domestic imperatives for luxurious and commodious living, inordinate life style and profligate consumption and production of wealth. And the millions of people unsettled from their land as a result of this will flee to the neighbouring countries for shelter and livelihood to find themselves enmeshed in group identity conflict with the receiving countries' native inhabitants. Thus, the western and rich North people's life style and their concept of development has come in for a greater security threat to the people of Bangladesh and in turn to India, when Bangladeshis will flee to neighbouring Indian states for shelter and livelihood. The northeastern states of India have turned into a burning cauldron of ethnic conflict due to the migration from Bangladesh induced by environmental degradation and scarcity of land. In the analysis of

Deudney while denying environmental security the status of national security is found lacking reference to these case studies with a view not to diluting the strength of his arguments. Even if his arguments that "most environmental problems are not international", "because", the causes, sources, problems, consequences and victims of environmental degradation are confined within the borders of one nation state[35] are accepted, as studied by Homer Dixon, environmental scarcity and degradation of resources within a territory can lead to various social effects in terms of poverty, economic deprivations, decline in agricultural productivity, migration and weakening of social and political institutions i.e. state which in turn contribute to intra state conflicts such as insurgencies, coups de tat, guerrilla warfare resulting in shifting of allegiance and legitimacy. These intrastate violent conflicts are having international impacts inviting intervention from external powers.[36]

Another distinguishing criteria Deudney applies to differentiate environmental security from national security is the intention involved. In the inter state wars or conflicts is clearly marked the intention of the parties involved, while mobilizing forces, procuring weapons with definite aims in mind. Since this is lacking in the case of environmental conflict, environmental security does not deserve the status of national security. Environment conflict that will arise out of degradation of resources between states through military means attest to the presence of intention involved in waging wars. That there is no explicit presence of intentional behaviour should not denude environmental security of its status being christened as national security. America's contribution to global warming is indisputably larger than other countries. Its impact in terms of sea level rise resulting in submergence of low-lying states such as Bangladesh, change of precipitation pattern and climate change is not the result of intentional behaviour of America to harm Bangladesh, India or others but certainly in the knowledge that its pattern of development, lifestyle and wealth amassing economy and limitless consumption contribute immensely to climate change. The rejection of Kyoto agreement by America which was designed to reduce emission of CO_2 and the hazardous effects of global warming on other nations clearly demonstrates its intention. The realist zero sum logic that "what is nuisance to others is gain to us", is applied. That America would reduce CO_2 on the plea that this will compromise the life style of its people, shows that America is aware of its repercussions on other nations and people. In this case is clearly ingrained the implied intention of the industrialized North powers in not reviewing what so far they have defined as development, universalization of which in the name of globalization has been a threat to the nature as well to the people dependent solely on it for their livelihood. Assume that there is no direct or explicit intention involved in the causation of environmental problems, to downplay its threat however severe these are, is illogical. This kind of absurd logic applied in international conferences and conventions on climate change and other important environmental issues while apportioning the burden and responsibility on powers for what have wrecked to the environment, would set a bad precedent and provide a scapegoat to scuttle the issue and pass on the buckle to others who happen to be the victims.

Deudney is of the view that "military organizations" which are "secretive, extremely hierarchical, and centralized, and typically deploy expensive highly specialized, and advanced technologies" to give protection from violence and threats are not the same with environmental security. " In contrast, responding to the environmental problems requires very different approaches and organizations".[37] Environmental degradation that rekindles intrastate or interstate violent conflict is not dissimilar with the other causes triggering armed conflict. With the advent of nuclear weapons, chemical and other biological weapons, the type of organization, methods and approaches required in case of classical military operations gets changed. Similarly, in case of environmental security the types of organization, techniques and approaches are bound to be different. If changes in techniques and approaches required in case of nuclear weapons do not in any way denude it of its importance, how the different approaches, methods and organization to deal with various aspects of environmental degradation could disrobe the environmental security of its significance and meaning? That environmental security needs different kind of organizations, approaches methods and techniques different from those of classical military warfare, according to Deudney creates "a conceptual muddle rather than the paradigm or world view shift".[38] Environmental security does include not all kinds of evils as threat but only those that are caused by environmental degradation or scarcity of resources along with its social effects which in turn creates intrastate violence with implications on international security and stability. The concept of environmental security is not a narrative analysis of the conflicts that arise from environmental degradation, but is suggestive of a new paradigm that challenges the established dominant ethnocentric view of development, realist notion of security and the divided hierarchical world and self vrs. other syndrome. It aims at a normative shift in this western worldview ubiquitously practiced as the bible of development – the sole cause of the despoliation of the earth and the destitution of the world. Ingrained in such a western view which is premised on Judeo-Christian tradition of anti nature, the critics of environmental security underrate the magnitude of the crisis, and challenges to the world view are considered as "dedefinition rather than a redefinition of security".[39]

To say that environmental security is a motivational, psychological and rhetorical strategy to divert the social, political and economic wherewithal now expended on military towards alleviating the environmental conditions is too narrow a view and negative approach. In the United States on many social issues 'war on drops', war on poverty or war on crime' movements as "moral equivalent to war"[40] have been launched in the past without any success. But these movements are different from environmental security in that the latter aims at reinventing the established view of the world, its political and economic thinking on which hinges the concept of national security. No doubt Ken Conca and others are right when they are of the view that discourse of national security has a set of powerful associations that cannot simply be redirected.[41] Because the discourse in national security is based on an established and western dominant view of the world which is very difficult to change. This is what environmental security aims at. The end of the cold war raised an apocalyptic ushering in of a new world order based on justice, equity, democratization of international institutions with multilateralism to replace the bipolarism. Nothing of the sort happened. Because everything depended on America and western powers, the key and dominant players crowning the world political system. It was America's unwillingness to cast off its dominance pontificating to the world its political and economic script to be followed in letter and sprit that could belie what Fukuyama to declare end of cold war as an epoch making 'end of history'. That does not mean the new and alternative perspectives on national security that came up even before and after the end of cold war have lost their significance. This is what led United Nations Secretary

General Kofi Annan to note in his Millennium Report that while security policy had traditionally focused on the defence of territory from external attack, it has now come to embrace "the protection of communities and individuals from internal violence". This focus on the protection of individuals rather than borders is the central element of what has become known as human security.[42]

Environmental security is more than a mere motivational or inspiring doctrine. It is based on a matter of fact understanding of the world phenomenon. It raises the basic question related to the established pattern of development, life style, the relationship between man and nature, the impact of the pursuance of such a type of development on future generation both living and non living, and the legitimacy based on ensuring inordinate life style, consumption and wealth production, which is taken for granted beyond space and time. It focuses on the security, development – governance nexus. Prescriptively, while not discounting the military response to crisis created directly by the environmental degradation and scarcity of resources, it stresses on non-coercive prevention policies – from preventive diplomacy, conflict management and resolution strategies to security system reform and the promotion of good governance, sustainable and equitable development, and democratization.

Environmental security is not synonymous with a motivational strategy. It is a normative and ethical paradigm aiming at awakening the people to the dangers of environmental crisis which is the result of persistence in the western pattern of development, industrialization, extravagant life style and inordinate consumption a synonym for heightened happiness and euphemism for being highly civilized. Are the people of America and North ready to come down to level of a poor man in South with the knowledge that this will not compromise the life style of their future generations? Accustomed to such an extravagant and commodious living, the people in North are not ready to awake themselves to such ethical imperatives required of them to ensure environmental security. That there will be reactions, discordant behaviour and opposition to such an awakening in spiritual terms, as a part of environmental security, does not divest environmental security of its meaning, and essence. Fundamental questions raised to review the inordinate life style and way of life for a sustainable world make environmental security a viable and alternative security concern to replace the traditional concept. If this be definition, scope and gamut of environmental security, then Jyrki Kakonen's fears of "militarized environment"[43] are ill founded. Because environmental security never advocates military prowess necessary to protect the people against threats from environmental degradation. As nuclear weapons make nuclearisation of war, the environment makes security the environmentalisation or greening of security.

It is also argued that war, nation and state are intrinsically connected and reinforced by each other in realism's frame of security. As the historian Michael Howard has observed : "self consciousness as a Nation implies, by definition , a sense of differentiation from other communities, and the most memorable incidents in the group memory usually are of conflict with, and triumph over, other communities. It is in fact very difficult to create national consciousness without a war".[44] Echoing the dominant perspective of national security strengthened by war, Charles Tilly observes that "states make war and war makes states."[45] These attributes of jingoism of war distinguishing nations, states and communities from other communities, nations and states and uniting one against the other in a divided and hierarchical world, it is said, are found absent in the concept of environmental security. Hence the lack of the cementing force of jingoism in environmental security divests it of any qualifications to be called as environmental security. Environmental security argues for a redefinition of national security not in the sense of one vrs. the other or uniting one nationality against another but in the sense of transcending the Westphalia model of security tied to the sovereign state system on the basis of self help in a supposed anarchic international system. Since most of the sources and causes of environmental problems which lay embedded in western pattern of development based on fossil fuel based industrialization, wasteful inordinate consumption, life style and production of wealth are global and transboundary, national security is to be defined with environment as a fundamental component not in a Westphalia and European sectarian language turned global in scope during the last centuries but in terms of an interconnected, not only between and among nations but also between the nature and man and between the present and future generations, living and non living. It is not the jingoistic forces of war but the envision of a universal self different from the self help, tied symbiotically with others living and nonliving in a timelessness frame beyond territorial exclusivity and spatial distancing of traditional nation state, and new planetary interests, ethics and wisdom that will define the national security. The movements for global governance of the environment, the rise of non-state actors like NGOs transcending nationalities and territorial boundaries and environmental regimes point to the inadequacy of westphalian sovereign state systems to ensure security to human beings. Nobody can refute the fact that to deal with environmental problems, their management and regulation entails the expansion of state capabilities but not on military terms. This requires coordinated efforts of science and technology, socio-political institutions, finance, various organizations and state machinery including the international community. Jason Clay has been quoted to point out that in many parts of developing countries, and particularly its tropical rain forests, states are dispossessing and destroying indigenous people whose claim to distinct national identity is high but whose potential for statehood is very low or non-existent.[46] This is what the proponents of environmental security are going to address not in terms of enhanced military capabilities but in terms of relooking at the western pattern of development based on fossil fuel technologies and industrialization.

It is the imposed western liberal market economy in the name of globalization, the entry of multinational corporations into the developing countries that is mostly responsible for deterioration and depletion in their environment. Environmental scarcity of sources along with their social effects in terms of poverty, economic deprivation and weakening of state capabilities has led to in these countries insurgencies, group violence questioning the very basis of legitimacy. The cases of banditry in rural Sind in Pakistan, the guerrilla war waged by the Sendero Luminoso in Peru and the New People' s Army (NPA) insurgency in the Philippines are good illustration of the links between environmental scarcities, grievance and violence.[48] What has led to coercive conservation[47] in many fragile and despotic regimes in African countries by dispossessing local inhabitants of their traditional natural resources, is not part of a design to fulfill the requirements of environmental security, it is to benefit state elites, multinational corporations and the western nations. The very argument that environmental security will lead to ecofascism seems preposterous and not in line with what environmental security is meant by. Environmental security and authoritarianism are antithetical to each other. The very stipulations of environmental security envisage the democratization of governance, participation of all people including the locals and indigenous, the non-state NGOs, in environmental conservation loosening of the Westphalia politico-military construct of security straitjacket, review of the western pattern of development, western life style, replacement of anthropocentric relation between man and nature and inculcation of new values, nuances, and new vision into the individuals' schema of life and planetary interests to determine the national interest of state. On the contrary, the western democratic regimes especially the United States seem to be democratic within, but in their relation to other countries in South for procuring natural resources and raw materials as part of their politico-military construct of national security, they have exhibited the signs of authoritarianism symptomatic of military regimes. The recent Gulf wars stand witness to this kind of authoritarian behaviour. In all international conferences and conventions whether on climate, forest, biodiversity, intellectual property rights or WTO the definition of national interests of rich industrialized North in terms of domestic imperatives such as inordinate consumption and exorbitant life style and production wealth, remained dominant. If Kyoto Protocol dealing with stabilizing of carbon dioxide in the atmosphere comes into effect, many developing countries will be carbon sinks for the developed world, resulting in many indigenous people being evacuated for creation of forests or trees cover as a part of climate trading system. Then what comes as a threat to the indigenous people is not the requirements of environmental security but the dominant developed world's definition of national security in terms of domestic imperatives such as production of wealth, ensuring exorbitant and inordinate life style and consumption, for which the large chunk of historic emission of CO_2 in the past and the present to the atmosphere of the developed countries can be accounted.[49]

Contemporary national security is closely connected to the institution of state sovereignty as defined in Westphalia convention. In the international realm, sovereignty has come to mean the existence in polity of a final and undivided authority over a particular territory, which only states can possess, and the reciprocal recognition of this authority which states extend to one another.[50]

Claimants within state and outside the state to the prerogative of sovereignty are considered not legitimate but encroachments. Similarly it is said, responding to international and global pulls and pressures exercised upon state sovereignty on environmental issues and problems mean a diminished or dilution of the sovereign state's legitimate authority. Thus, the requirements of environmental security come in direct clash with the enhanced concern for its sovereign prerogatives. When nations clash over environmental scarcity of renewable resources such as river water as found in various parts of the globe, for example, India and Bangladesh over Ganges water, water conflict in Middle East and Africa they lay claims to water within the nonnegotiable prerogative of sovereignty. When intrastate conflicts in terms of group violence, insurgencies, and guerilla war fare occur in fragile and weak states presided over by authoritarian regimes and exacerbated by environmental scarcity of resources in combination with other factors like social, political, economic and cultural context, these conflicts occur over the issue of legitimacy to the sovereign state system. The environmental issues which are global need global governance and management. In that case, traversing beyond the straitjacket of sovereignty is what is required of environmental security. Traversing beyond is not always an erosion of sovereignty rather an extension of its effectiveness and influence in dealing with environmental problems. Changes in international situation are taking place so fast and pervasively that to think solution in terms of the rigid logic of Westphalia sovereign state system in the changing scenario is an expression of European arrogance. Barry Buzan who recognizing environmental degradation as a security threat keeps it at par with economic and political factors, is of the view that "the sources of threat are also diversifying away from the state. Many of the new threats seem to stem from complex systems both natural (the ecosystem) and the human made (the global economy), and the operation of these systems is often poorly understood."[51]

Two perspectives on sovereignty and environment linkage continue to be debated. The first perspective led by Deudney and others observe that there has been an erosion or weakening of sovereignty. Environmental concerns are said to be erecting new and effectively global standards for state behaviour. These new global standards are said to manifest themselves in many ways : in formal dealings among states such as the creation of international environmental regimes, in rules of environmental conditionality attached to the actions of international organizations such as the World Bank[52], in the evolving norms of a growing body of international environmental law[53], and in the political pressures brought to bear on governments by increasingly transnational environmental movements, citizens networks, and non-governmental organizations.[54] Ecosystems and environmental processes do not respect state borders, sovereignty itself becomes a key institution of global scale environmental destruction. It creates a scale for decision making, adjudication, and authority that does not coincide with fundamental ecological realities, and thus frustrates ecologically responsive management.[55]

The second perspective is that the emergence of multilateral institutions for environmental protection, donot inevitably erode state soverignity and may even strengthen it. By placing states at the center of institutional responses and strengthening their capacity to act collectively, it is argued the menu of choices available to state is being expanded not restricted.[56] Levy, Keohane and Haas have argued that, although environmental regimes may limit the scope of governments to act unilaterally, they also facilitate collective state based problem solving.[57] Ken Conca striking a balance between the two perspectives states that "sovereignty is in fact being transformed as a result of global ecological interdependence, but not in the manner sketched by either of the above claims, or even by the net effect of the two taken together."[58] A new universal norm of environmental responsibility is emerging in a highly segmented set of activities, including the lobbying of scientists, the pressure of public opinion, the calculation of governments, and the targeted political pressures of eco activists. Responding to international environmental pressures can create resources and purchase legitimacy at the same time it may constrain the menu of policy choices. Sovereignty as conceptualized in Westphalia carries with a complex bundles of rights : equality among states, non-intervention, exclusive territorial jurisdiction, the presumption of state competence, restrictions on binding adjudication without consent, exclusive right to wield violence, and the embeddeness of international law in the free will of states.[59] This is not to argue that a particular set of international pressures affects these various components and norms of sovereignty equally or in parallel fashion. There is reason to expect that some normative pillars of sovereignty can be strengthened as others are undermined or eroded. For example, in the case of transboundary pollutant flows, "institutional mechanisms" and agreements "to control them could erode the sovereign right to exclusive territorial jurisdiction, but at the same time strengthen aspects of the principle of non-intervention, if the flows themselves are viewed as unjustified intervention".[60] The Westphalian straitjacket sovereignty is based on the legal framework that it has exclusive jurisdiction over the territory. This constitutes the raison de tre to claim the status of the international standing. The claim of the enduring sovereignty in the face of environmental pressures stresses states as problem solvers. Deudney's views that "an enhanced state concern for its sovereign prerogatives could greatly impede international environmental cooperation[61], stand discredited. Hardly the ink on the major agreements signed at the 1992 UNCED conference at Rio dried than the Brazilian diplomat Marcos Azambuja offered the following analysis ;

"Brazilian interests are reinforced in the majority of the documents. At no time did we face opposition to our basic interests ... We came out of the negotiations without the slightest scratch to our sovereignty".[62]

While assuming that the concept 'sovereignty' is based on the foundation of territorial exclusivity and spatial distancing, its internal capability to ensure that the legitimacy given to it by society and various sections of people on the basis of its performance continues, is lost sight of by those who want to confine environmental security to the Westphalia procrustean bed. Environmental scarcity and degradation of resources when coalesced with other economic, political and cultural factors accentuates economic deprivation, decline in agricultural productivity and weakening of political institutions such as state. In this situation "if the state fails to meet the grievances of the disadvantaged people, then threat of violence, guerrilla warfare and other forms of conflict manifest challenging the legitimacy of those who are in power and their claim to sovereignty. This internal failure in terms of lack of capability and effectiveness to ensure environmental security is reflection of limitations of sovereignty. When Brazilian government was very much concerned that there are no scratches on the wall of its sovereignty, its inside surface shows signs of being slightly scratched. The following excerpt shows the irony:

Whenever one looks in the Amazonian economy, the state is in retreat; unable to finance tax breaks or build highways without the aid of multilateral banks, unable to include more than one percent of the rural

population in official colonization schemes, unable to control land tilting or land conflicts, unable to register or tax the greater part of the Amazonian economy, unable to enforce federal law on more than a sporadic basis.[63]

Efforts to scratch the wall by constructing a regime for the preservation of the world's remaining rainforests and of the Amazon in particular were defeated. Assessment of the limits of state capabilities comes at a time when the Brazilian state has been placed squarely at the center of most schemes for sustainable development in the region. Far from prohibiting state action, the net effect of international pressures has been to stimulate a more active, interventionist role in the region under the rubric of supporting sustainable development. Without eroding sovereignty, these pressures strengthened the presence of state in the region and in Brazilian society as a whole. A multifaceted character of sovereignty is emerging where " the freedom of state to undertake, promote or tolerate process of environmental degradation is being limited and many of the limits emanate from sources external to state itself. At the same time there is little doubt that new international institutions have made some governments more effective problem solvers. That both effects could be happening at once is testimony to the multifaceted character of sovereignty."[64] What is lost sight of by the many critics of environmental security is that the internal forces questioning the capability of state to deal with environmental scarcity of resources are downplayed to magnify the international environmental agreements as constraints on external sovereignty. But as seen from the above analysis, rather the international environmental forces and pressures strengthen the role of the sovereign state system as problem solvers. The requirements of environmental security do not aim at the proselytization of sovereign state system or its erosion. It aims at a coordinated and cooperative efforts of sovereign state systems in dealing with the management of environmental problems. Given the magnitude of the crisis and the threats to the very survival of humanity on the earth, the new planetary interests would constitute the national interests of sovereign states. The need is to reflect upon what this means for humanity as a whole, and for each nation state. It is time that the planetary interests became the principal political concept guiding decision-making on survival issues such as sustainability and environmental integrity. It is time that the sovereign nations pursue a legitimate national interest that is in each case, compatible with the greater good of us all. On the true global issues, a new approach for decision-making is required. Instead of negotiations being the consequences of competing claims of all national interests, the global objectives need to be identified first and accepted as binding, with subsequent negotiation determining what national obligations must arise from this. The relationship between national and planetary interests has been the focus of environmental security. The impending environmental crisis global and transboundary in its nature and scope coming as greater threats to humanity, calls for the institutionalization of supranational loyalties to the planet and to humanity as a species. This means a subordination of national sovereignty to the imperatives of environmental ethics and values.[65]

In 1995 a commission headed by former Swedish Prime Minister Inngvar Carlsson and former Commonwealth Secretary General Shridath Ramphal declared : "the idea that people have common interests irrespective of their national or other identities and that they are coming together in an organized way across borders is of increasing relevance to global governance".[66]

Alleviating the fears of eco fascism of the critics of environmental security, the very requirements of environmental security suggest a global governance of environmental crisis with involvement of all institutions including the state, non-state actors, regions and groups. Given the sources of many of the new threats stemming from complex systems both natural and human made, and diversifying away from the state into diverse, regional and local in character, the monolithic statist Westphalia approach quite familiar with realist paradigm seems to be inadequate to solve the environmental crisis. "Global environmental governance emerges because the spatial scale of the state is inadequate in dealing with the scales of environmental change, and thus the practices of governance move towards regional and global levels and at the same time towards local levels in response."[67] The ecological project of governance envisions a world of small interacting communities, what Bookchin named as municipal confederalism[68] or what Karliner termed it as grass roots globalization,[69] which means a myriad locally organized, but transnationally networked, social movements and NGOs are becoming increasingly mobilized to resist the dominant practices of TNCs and states. To create alternative forms of the western dominated political economy the most common phrase used is the "common". It is defined as a site of ecologically sustainable and socially just political economy. This means the construction of a myriad small-scale societies organized around common regimes through resistance to the globalising practices of TNCs, state and international development agencies.[70] To quote Caldwell :

Rhetorical assertions of national sovereignty continue to be heard, but the imperatives of geophysical hazards to all nations are pushing their governments toward modification of their asserted freedom to act as they please in relation to their natural resources, industrial practices, and the environment. A rhetorical strategy to evade the sovereignty road block is the concept of 'merged sovereignty', which infact occurs when nations seek to realize there own national objectives through treaties with other nations having interests in common.[71]

Deudney is of the view that "national identity and security collides directly with world views and identities supportive of sustainable environmental practices." The nation is not an empty vessel or blank slate waiting to be filled or scripted but is instead profoundly linked to us vrs. them thinking". Further he states that "nationalism means a sense of us versus them, of the insider vs. the outsider, of the compatriot vs. the alien."[72] The "us vs. them" thinking comes in direct conflict with environmental political thought which stresses on 'one world' and global village sensibility of environmental awareness.

As Homer Dixon and his associates have shown in their studies, many conflicts that arise directly out of environmental scarcity of the renewable resources such as water, fish and others can be put into the "us vs. them" framework, where one nation knows that the causes of conflict lay directly with the environmental degradation of certain sources of the other countries. Where there is differential degree of contribution among nations to the degradation of the environment, lack of scientific precision to quantify the magnitude of each country's contribution, and scientific certainty about the imminence of environmental catastrophe and about its country and region wise impact, all argumentations against applying this "us vs. them" paradigm into international relation on environmental problems seem to be contextualization of national security within the Westphalian model. On environmental issues in international forums a clear line is drawn between North and South. The seemingly inapplicability of this "us vs them" thinking does not mean that environmental crisis is not a national security issue in the sense of a realist framework but in the sense that the environmental problems will affect the livelihood of its citizens whose security has traditionally been tied with territorial impregnability. How security defined in terms of this territorial impregnability, exclusivity and spatial distancing stands in the face of flow of pollutants, climate change, change in precipitation patterns, global warming, acid rain, ozone hole impact? The scientifically arrived at findings can no longer be ignored in the name of adherence to classical model of sovereignty and Eurocentric realist understanding of the world and on the other hand requires rethinking of security in terms of environment.

This reality of struggle among nations for survival and dominance resulting in elitism enforces local and domestic hierarchies and legitimizes the existence and use of military apparatus, while life is permanently threatened by the imperatives of military security. Military imperatives dictate that states develop and deploy the most effective military technology available. The effects on land and nature are little considered. No military technology could be more environmentally damaging than nuclear weapons. Thus the environmental effects of producing, storing, developing, and dismantling them not to mention the effects they would have on the environment if ever used, are considered secondary. A similar argument could be adduced about the entire range of military technologies, from cluster bombs to napalon to defoliants to biological weapons.[73] Land has no moral or little worth except that it is a spiritless other that humans can rightfully exploit to serve their own ends. Just-war theory generally evaluates collateral damage's significance in the context of civilians killed or injured due to military operations. Yet collateral damage also kills and injures animals and destroys biotic nature that have even less stake and less say in the conflict than civilians. Where is their right to life ? The realist ethic to disrupt or destroy the evolutionary process which includes all its organic and non-organic components, reducing the diversity of life and stability and beauty of the natural system is unethical.[74] Responding to power politics, the ecologists allege that if power politics is real, it is a constructed and deconstructable reality. "Reality is a social construction". It is "created and constructed by beliefs and behaviour. Structures do in fact shape beliefs and behaviour the way some positivists thought, but these structures are the product of human action".[75] To hold such logic of structural realist assumptions of power politics as permanent is the result of application of Thomas Hobbes and Issac Newtonian view of man "as particles of constant motions" to social relations.

From the realist axioms of structural and motivational conflictuality flow a series of corollaries with a devastating impact on life and freedom, of which hierarchy or elitism is the obvious problem. This gives rise to the emergence of a strong state always embarking on devising a system of accumulation and control from the top.

Realism is also criticized for its homogenizing worldview in its power politics framework. It reduces motives to a few universals, such as power quests and physical growth. "In the process, realism has also homogenized the political unit by emphasizing a statist ontology that not only ascribes a natural aura to such a form of association, but also minimizes its internal variance".[76] This is particularly characteristic of Waltz's third image.[77] By reducing all to a narrow groove of power motives, and holding these as immutable- and permanent, realism imposes a structural straitjacket upon all cultural issues, diversities and anthropological insight. In the words of Eric Lafarriere, "From an ecological perspective, the power assumptions and political language of realism can only portray the world as a giant machine fixated on the goal of survival or victory; they can only trivialize culture and accultured individuals in the process of social creation".[78]

In its idioms of anti ecological dimension of power, realism depicts the world as basically materialistic, conflictual and elitist devoid of any social utopia and idealism.

By arraigning one individualized self in case of a group or class, or one super power nation, against the other to satisfy the elite's interests in terms of accumulations, consumption and control realism super imposes an artificial, fictitious construct over and above a universalized or expanded self, typical of ecological portrayal of the natural world. "By limiting the definition of threat to the other's potential ambitions over the finite, zero sum game of the state system, and by turning this preeminent reality into the main engine of history, the call to arms by realism turns nature into raw materials and legitimizes warfare as a vehicle of the good".[79]

Impelled by the objectifying bias of Newtonian science, the positivism and behaviouralism taking cue from the classical realism took the recurrent pattern of history as constants, quantified the power and state capabilities and replaced ethics with problem solving. Thus, realism as the ahistorical science of the state stunting human development and questioning its survival stood opposed to organicism, subjectivity and historicism characterizing the ecological thought. Critical Theory challenges this statist denominations of realism. "If states dominate the arena, this is a feat of power politics repressing other dimensions of reality that could potentially replace the states if an emancipatory praxis could- with the help of Critical Theory - empower other subjectivities than those who dominate at present. The social world does not exhibit any iron laws, all regularities can be broken, and it is the task of Critical Theory to show this...".[80] Ecological approach can add to completion of emancipatory project in this direction by relying on what Bookchin said earlier about municipal confederalism. Critical Theory can take non-military environmental security threats as a part of its argumentation against established discourses of security, thereby relativising conventional wisdom.

Critique of Liberalism

Liberalism which started as a doctrine of freedom was based on possessive individualism of Hobbes and Locke in search for accumulation of wealth, profligate and commodious lifestyle. At the individual level, he/she needs freedom to achieve the goals set by the materialistic world opened before him/her by the Newtonian age. At the state level, liberal democratic government legitimized by the individuals' criteria of wealth production, profligate and inordinate lifestyle and commodious living emerged as the guarantor of these goals to its individuals by introducing into the world, its market economy, liberal ideas, free exchange, trade, and protectionism.

The liberal tradition set by America and Western allies was to guarantee these objectives to the individuals at home by getting unimpeded access to resources available any where in the world even by the application of force. Like realism, it viewed the nature and other nations as the other to be exploited for satisfaction of individuals' material needs at home. The liberal institutionalism cannot escape its entanglement with and commitment to the capitalist market system. Its anthropocentric view of nature and universe and consequent emphasis on economic growth leads to a relative lack of concern for the stability, integrity, and beauty of the earth. Obsessed with growth fetish the western liberal market economy destroyed both the nature and the poor. In other words, "If Judeo-Christian monotheism took nature out of religion, Anglo-American economists took nature out of economics"[81]. Overconsumption and inordinate life style of the western people has proved to be a nuisance to nature. The liberal tradition combining individual consciousness with responsibility is eminently attractive to ecologists. Liberal literatures in International Relation articulate a normative commitment to freedom by favouring a cooperative utopia based on open exchange, material growth, positive law and specialized knowledge. These include the early advocates of free trade,[82] the functionalist and neo-functionalist arguments stressing the relationship between peace and elite co- operations[83] the regime literature and its descriptive emphasis on 'mechanics of cooperation,[84] and part of the contemporary normative literature on global humanism.[85]

The liberal cosmopolitan project aims at the attainment of a global commonality with the gradual eliminations of borders. But the overriding concern was unity. From an ecological perspective, the question pertains to the purpose of liberation in espousing unity. The ecologists are in favour of global concert but surely not at the cost of integrity, and diversity of member communities. The worst nightmare for the ecologists is the current spread of western liberal market economy throughout the world culminating in globalization. It involves the most fundamental centralized restructuring of socio-economic and political relations since the industrial revolution. It has created a new corporate colonialism in TNCs to which all individuals and nations must adjust. The new corporate colonialism is likely to dispossess, impoverish, and marginalize more people, destroy more cultures, and cause more environmental devastations than either the colonialism of old or the development of the last 50 years.[86] In the current context it means the homogenization of prices, products, wages, rates of interest and profits to become the same all over reinforcing the economic interests of North countries. The process of economic globalization advanced by the Bretton Woods institutions promoted the interests of the superrich so successfully that it also damaged their concern for and obligations to less fortunate inhabitants and their natural ecosystem in the developing world. In other words, it is nothing but another euphemism for the continuing exploitation of the underdeveloped by the developed, and redistribution of resources and wealth from the less to the more developed.[87] Globalization has not created a global village in transcendence of national boundaries, rather a highly divided, hierarchical elitist political structure. Thus, the ecologists' critique of liberal cosmopolitanism stands vindicated in that the homogenizing logic of liberalism is ruining ecosystems and decimating cultural specificity. The integration theory of the 1950s and 1960s for all its neutral tone, also assumed a basic link between global freedom and increasing communication.[88]

In contemporary liberalism, as held by James Rosenau, global culture based on 'reason' and 'analytical skills' can be disappointingly defended as an emancipatory tool for the individual.[89] Trade based globalization unfolding of western liberal market based economic-political order has been raised to a high moral level a synonym for development to which the developing world has been asked to dedicate themselves in slavish adulation of what the west terms it as good. Siphoning off the resources and materials from the developing world like an inverted blood transfusion from the sick to the healthy, the globalization has enabled the rich North to promote their domestic imperatives in terms of guaranteeing the profligate, inordinate consumption, life style, and production of wealth. Thus the approach of liberals to global growth is based on total disregard of the imperatives of finiteness and diversity of nature, and the ever increasing hiatus between the rich and poor, freedom of the individuals and right to life of all other non human living and non-living.

The ecological critique should examine what David Mitrany and Harold Laski propounded their commitment to the materialist utopia in their defense of efficient servicing. For Mitrany, servicing formed a clear path to freedom.[90] Normative endorsement of growth is also found in Peace studies and within the Southern Perspectives on world order reform.[91] Ecologists can appreciate Mitrany's attempt at replacing a mere voting democracy with a working one.[92] But they are distrust of "the enlightened mission of experts". In their views, "technocratic depoliticisation, rather is viewed as the embodiment of totalitarianism, the vehicle for a very distinct brand of politics".[93]

Ecologicalising Emancipatory Mission

An ecological approach to International Relation Theory and global governance rests on alternative ontologies, epistemologies, and ethical/

political prescription. The mechanical view of nature and attribution of the 'other' needs to be reconsidered. Nature is finite, whole, an interdependent body connecting all living matter in a cyclical and non-hierarchical manner. It is more often a function of cooperation within and between species. Diversity and timelessness are the other attributes of nature. Diversity allows nature to withstand the demise of individual life forms while maintains its grandeur, beauty and bounty. Timelessness shows life is not imposed or spontaneously created but is undissociable from time. It is a long, slow and cumulative process. Recognition of longevity supports the assumption of mutual aid in nature. Thus it rejects the traditional utilitarianism, and mechanical view of nature, and ontologies of conflict. What Pope John Paul II said in the Conference of Roman Catholic bishops in the Philippines and Dominican Republic in 1988 deserves excerption : "... One cannot use with impunity the different categories of beings, whether living or inanimate – animals, plants and natural elements-simply as one wishes, according to one's economic needs. On the contrary, one must take into account the nature of each being and its mutual connection to an ordered systems, which is precisely the cosmos".[94] It advocates a harmonization with physical surroundings admonishing against imposition of structures of domination. The principle of classical theory of democracy, freedom, individual fulfillment and equal treatment are vindicated by ecological ethics. Instead of looking the world of nature as the other, a more coordinated view of the world embracing both the world of nature and the world of humanity[95] is what the ecological approaches insist on. "To deny moral value to ecosystem because they are not integrated in the same way as centralized individual organisms is to make a category mistake. In thinking about the value of social and biotic communities we should be concerned for a matrix of interconnections between centres, not for a single centre".[96]

Thus the mechanical view of the world and nature must be replaced by an organicistic one[97] which reads the world not as a soulless machine subjected to cause and effect, and repetitive patterns – sine qua non of scientific investigations, positivism and behaviouralism. To confine the study to the limitation of cause and effect syndrome means to fall short of understanding the complexity of nature and its inextricable connections with the evolution of life in a timelessness frame. Not that these scientific investigations, assumptions and methods are not helpful, but should not constitute the principal means of understanding or attributing meaning without organicism being attached with mysticism.[98] The ecological project calls for a rejection of the detached object of study, objectivity and mechanism forming the core of the Cartesian – Newtonian dominance in social sciences in favour of subjectivity.

The ecological teleology points to the significance of historicity and dialecticism in understanding the current human conditions as the dynamic ebb and flow and product of synergistic forces through time. This means to get rid of Westphalian straightjacket [99] the prevalence of ahistorical bias in formulating the concept of international system reinforced by the dominance of an economistic, natural science based understanding of the social world.

"A silent acceptance of this cultural bias" the imposed dominant economic and political thinking of western paradigm— "is virtually universal, not only within the context of a western centred parochialism, but also among IR scholars who have tragically abandoned their own quite different historical and cultural traditions".[100] Seducing all cultural diversities, civilisational essences, regional specificities and bioregimes' particularities and their coordinated interconnections woven into the warf and woof of cosmos, into wearing the western straitjacket is to block communication of the mainstream International Relations with other disciplines and snap man's eternal chord with the cosmos. It calls for a multi disciplinary and multidimensional approach that will travel beyond microeconomic analogies, handling forces that are much more elusive and much less amenable to behavioural manipulation than the heavily magnified state. Awakening it from the Wesphalian procrustean bed, it should be studied as an historical action and a social actor among many. Thus, the ecological vision includes a system cosmology, a theory of democracy of global structure of governance, a new scientific epistemology supplementing behaviourlism and Newtonian objectivity, a strategy for global peace and security on the basis of cooperation and a recognition of intergenerational equity and accountability, and interspecies responsibility.

Revisiting Globalist Criticism

Deudney is impervious to the above changing views on the relationship between man and nature and its impact on the international relation. Critique of realism and liberalism from an environmental perspective challenges the very basis of Deudney's arguments denouncing the linkage between environmental degradation and conflict. To sum up, Deudney's globalist perspective sees environmental politics as subversive of the state, and the state as subversive of the emergent global environmental political sensibility. He makes three basic claims. First he claims that it is 'analytically misleading' to characterize environmental degradation as a threat to national security, because "the traditional focus of national security has little in common with either environmental problems or solutions". Second, he argues that "the effort to harness the emotive power of nationalism to help mobilize environmental awareness and action may prove counter productive by undermining globalist political sensibility. Taking the position that the security notion is still primarily linked to the states system and national security, he argues that "environmental degradation is not a threat to national security. Rather environmentalism is a threat to 'national security' mindsets and institutions". He further buttresses that dressing the environmental program in the "blood soaked garments of the war system" will undermine environmental "core values" and create "confusion about the real task at hand". Third, he contests the thesis that environmental degradation will be a significant cause of interstate wars.[101] Eric K. Stern disagreeing with Deudney is of the view that his "globalist orientation actually undermines the foundations of the domestic international decision. Recent history, most notably the tragic civil wars in Somalia, and Rwanda, and the conflicts associated with the break up of the former Yogoslovakia, indicate that this distinction is becoming less significant to the international community."[102] Buzan's conceptualization of security, which challenges the privileged status of national or state security, provides a more realist approach. He observes that "the concept of security binds together individuals, states, and the international system so closely that demands to be treated in a holistic perspective, and that "a full understanding of each can only be gained if it is related to the other too". Given this, "attempts to treat security, on any single level invite serious distortions of perspective".[103] While 'balancing enduring tensions between the several levels at once', Stern is of the view that, "an increasing sensibility to environmental threats by national communities and a resulting incorporation of environmental values into national security policy does not necessarily undermine more global efforts to protect the planetary environment".[104] He is of the further view that "the likely scenario for the next half century is one of a mixed system of governance characterized by an uneasy distribution of authority between autonomous sub national entities, national governments, and emergent supra national institutions, such as regional and global formal organizations, regimes, and a developing body of international law".[105] The post cold war period resembles to that of medieval Europe with its characteristic uneven process of integration and disintegration.[106] Stern also questions Deudney's characterization of common security as a marginal phenomenon and an unstable conceptual foundation for environmental security. 'By equating security exclusively with the national security from violence problematic, Deudney "tends to discount the significance of a broad range of parallels and interrelationship between environmental values and military security". He further states that Deudney's "approach prematurely dismisses the viability and persuasiveness of conceptual innovations building upon the notion of security and acknowledging the interdependence that actually characterize the problem of 'security from violence' to nearly the same extent as the problem of protecting the global environment".[107]

A Critique of Orthodox Realism

Rejection of treating environmental problems as security issues with a narrow definition of national or international security delimited to concern with threats originating in the military capabilities of the other states has been the main thrust of mainstream contemporary American realists. Reflecting the view held by mainstream American neo realism, Stephen Walt has recently defined the scope of security studies as "the study of the threat, use and control of military forces".[108] He argues that the security studies 'explore the conditions that make the use of force more likely, the ways the use of force affects individuals, states, and societies, and the specific policies states adopt in order to prepare for, prevent, or engage in war".[109] According to Walt, incorporating non-military phenomena expands the scope of security studies. In his words, "defining the field in his way would destroy its intellectual coherence and make it mere difficult to devise solutions to any of these important problems".[110] "Walt's realist orientation is an orthodoxy seeking to protect a core of security studies focused on state and military security issues".[111]

Adherence to this conceptual exclusiveness- does not prevent Walt from incorporating "economics and security" in a peremptory place in his proposed "research agenda for security studies".[112] He justified his stand by pointing to the linkage between military spending and economic performance, posture resources, and the impact of the military industrial complex. But his exclusion of other non-military threats from his security agenda is less understandable. "once the door is opened to non-military economic activities, it becomes difficult to justify exclusively other non-military issues from the security studies agenda".[113] A study of the links between environmental degradation and military issues in terms of environmental contamination from weapons development / production, pollution and influence attempts, the environmental consequences of warfare and environmental degradation as a cause of conflict will show the unjustifiability of Walt's exclusion. The environmental consequences of production, storage and use of nuclear weapons, biological, and chemical weapons, and even the conventional weapons development, production, and storage[114] have led to discard the non admission of the linkage between environment and security on the part of realists. Deliberate pollution has been a significant military and political device designed to influence the behaviour of other actors in world politics.[115] The intentionally taking steps to produce environmentally catastrophic consequences in pursuit of political aims by the terrorist and non state actors being enraged at the injustice meted out to them in world system, cannot deny the linkage between environment and security. According to T.W. Hawley, the oil spills and oil fires of the Gulf war unleashed by Iraqi forces in 1990-1991 were "an environmental catastrophe on the order of the explosion at the Chernobyl plant or the lethal release of methyl isocyante at Union Carbide's chemical plants in Bhopal, India".[116]

"The threat and intentional inducement of environmental catastrophes bears a striking family resemblance to the threat and use of economic sanctions as a diplomatic instrument, the importance of which Walt concedes".[117] Examples are not wanting to show the intentional inducement of environmental disasters as a part of warfare is not a new phenomenon. It has already been pointed out by Dixon that environmental degradation along with its social effects causes simple scarcity conflicts, group identity conflicts, and relative deprivation conflict which increase the probability of resort to inter and intra-state violence. Other analysts also similarly argued that the economic and social dislocations associated with climate change may result in civil and international wars,[118] Lodgaard's proposal for a regime of environmental confidence and security building measures (CSBMs) to mitigate the tendencies to violence identified by Homer Dixon further points to the untenability of a rigid separation between the environmental and traditional security concerns.

Another interface concerns opportunity costs,[119] multiple purpose capabilities and conversion. Opportunity costs are incurred when assets invested in military security projects become unavailable for the purpose, a phenomenon Buzan sees as a part of "defense dilemma".[120] Such tradeoffs occur between defence and education, health care and environmental clean up. Another interface is the multiple purpose capabilities which the military through its capability is engaged in various rescue-and disaster relief missions required of in natural cataclysms or technologies disasters having serious environmental consequences. Finally, conversion alludes to the possibilities of reallocating existing military assets and capabilities to other uses outside the military sector such as environmental clean up measures.[121]

Stern sums up saying the realist argumentations against incorporating the environmental phenomena into the security agenda are not well founded.[122]

Reformist Criticism of Environmental Security Revisited

Levy defines the term environment as referring to "issues involving biological or physical systems characterized by significant ecological feedbacks or by their importance to the sustenance of human life".[123] He makes a commendable attempt to specify the central terms involved and to assess soberly the costs and benefits associated with integrating environment and security. Stern, on the other hand specifies the types of environmental values most likely to be perceived as security threats in the following, admittedly anthropocentric terms: "threats to human health (such as highly contagious bacterial or viral diseases), threats to essential resources which support life in human eco system (e.g. air, water supply, and food production), and threats to the integrity of valued non human ecosystem and local / global biological diversity".[124] Three arguments made by Levy deserves examination. First, Levy holds that "the assertion that many environmental problems constitute security risks is correct, and is of very little importance".[125] He argues that the two key threats – ozone depletion and climate change-do appear to constitute "environmental problems that currently pose a direct physical harm to US interests", but he questions whether anything is gained by labeling them security issue.[126] Second, Levy argues that taking these environmental issues out of the realm of "low politics" where significant progress has been made may actually be counterproductive. Third, Levy recommends that the study of indirect threats produced by environmental induced conflict be left to researchers studying more general processes of social conflict, instability and war. These researchers should incorporate environmental factors into their more general models of conflict escalation and de-escalation. Overall, Levy essentially defends continued compartmentalization and advocates more research within the traditionally defined sub fields of international relations and comparative environmental policy. Despite Levy's arguments giving a number of good observations on environment security his study is stymied by several serious limitations. First, walking on the same furrow left by Deudney, Levy deliberately puts a primacy on the issue of national security and was averse to considering the implications of tying the environmental issues with supranational security concepts. Furthermore, Levy seems to frame the more general types of conceptual issues on the basis of the specific configuration of US interests. That means his study of security and seriousness of the climate change threat is rated high and low on the basis of the magnitude of the disastrous effects it will have on US interests, natural resources and its capability to cope with these changes. The crucible is the US interests on which is to be tested the magnitude of the environmental crisis. On the basis of the impact on US interests, resources and capability, the qualifications required of environmental security to be incorporated into the mainstream of national security, are to be defined. If the scientists, prediction about the immediacy of effects of climate change on the survival of human lifes, more specifically the submersion of small island states under sea come true due to even the modest rise of sea level, then there are obvious reasons to argue for associating environmental security with the fundamental national security goal of survival.

According to Stern, problematic is with Levy's claim that it is counter-productive to raise environmental issue to the status of "high politics".

Levy challenges the assumption that assigning an issue to the realm of high politics automatically increases the likelihood that forceful action will be taken. But his argument hinges primarily on a single historical example–policy with regard to protecting the ozone layer. Levy claims that progress was made on this issue by "sub cabinet officials operating out of the limelight and with little Congressional meddling".[128] He claims that " it is hard to escape the conclusion that we probably saved more lives by treating the stratosphere as low politics than as high politics".[129] "Yet is it really possible to draw far reaching conclusions on the basis of a single case of alleged policy success ? It may be the case that other issues, arising under other circumstances might well be more effectively and rapidly addressed by a 'high politics' route. Stern further says that "issues requiring significant investment and research, and public sacrifice, might require high level and more public leadership."[130] Levy argues that climate change, in contrast to the relatively straight forward ozone problem is "much more like the problem of containing Soviet Union ; it requires a ground strategy to guide actions in the face of distant, uncertain threats, and an overarching commitment from high levels of leadership to stay the course through the ebb and flows of popular sentiment."[131] Thus, for this important environmental issue, Levy calls for a high politics type solution. Third, most importantly and seriously, he delimits his concern to only "links from process of environmental degradation to deterioration in security positions and excludes from considerations "connections that run in the opposite direction (from use of force to deterioration of environmental quality)."[132] In the words of Stern, " Nor are links between preparations for the use of force and environmental degradation are considered. Thus his overall conclusion is based upon an examination of only part of the case made by advocates of environmental security."[133]

Comprehensive Security

As seen from the above analysis, the security problematic is caught between two extreme positions. The traditionalist position represented by Deudney, Stephen Walt, and others envisages to keep environmental affairs off the security for variety of different reasons, none of which seems compelling. Stephen Walt giving probably the strongest statement of the traditionalist position is averse to widening the security agenda outside this strictly military domain. He argues that this tendency to widen the security domain "runs the risk of expanding 'Security Studies' excessively; by this logic, issues such as pollution, diseases, child abuse, or economic recessions could all be viewed as threats to 'security'. Defining the field in this way would destroy its intellectual coherence and make it more difficult to devise solutions to any of these important problems."[134]

Walt does import economics and security into his picture but only as they relate to military issues, and not as economic security per se. According to Barry Buzan, given the political function of the word security, it extends the call for state mobilization to a wide range of issues. Central among the arguments for the conceptualization of environmental security is the mobilization potential : the concept of national security

"has an enormous power as an instrument of social and political mobilization," and therefore " the obvious reason for putting environmental issues into the security agenda is the possible magnitude of threats posed, and the need to mobilize urgent and unprecedented responses to them. The security label is a useful way both of signaling danger and setting priority, and for this reason alone it is likely to persist in the environmental debates."[135]

So the traditionalists' criticism of the wideners that they risk intellectual coherence can be a powerful point. Further he said that "the wider agenda tends, often unthinkingly, to elevate security into a kind of universal good thing-the desired conditions towards which all relations should be moved."[136] The debate over whether security label is worthwhile appending to environment remains dominant in the conception 'comprehensive security'. It points to the emerging values of environmentalism establishing their own moral basis, which according to Buzan are already emerging as a new norm of international society [137] Deudney in a similar vein talks about ecological awareness being linked to "a powerful set of values and symbols" that "draw upon basic human desires and aspirations" which should be mobilization basis and not the regressive logic.[138]

Given the nature and scope of environmental problems which are global in character, the state centric paradigm of national security evades the solution. The concept of 'security' tends to carry with it the implications that the answer, the defence, is to be given by the state. This is what stated by Moss :

"The most serious consequences of thinking of global change and other environmental problems as threats to security is that the sorts of centralized governmental responses by powerful and autonomous state organizations that are appropriate for security threats are inappropriate for addressing most environmental problems. When one is reacting to the threat of organized external violence, military and intelligence institutions are empowered to take the measures required to repel the threat. By this same logic, when responding to environmental threats, response by centralized regulatory agencies would seem to be logical. Unfortunately, in most cases this sort of response is not the most efficient or effective way of addressing environmental problems, particularly those that have a global character."[139]

As argued above, the concept of security includes the message that here is an issue where the state is or should be licensed to take appropriate steps, but " the instinct for centralized state response to security threats is highly inappropriate for responding effectively to global environmental problems".[140]

At the other extreme is found the concern for environment in a privileged position, in effect tumbling the threat of international violence from its traditionally privilege and hegemonic position at the core of the security discourse and replacing it with a new environmental "king of mountain".[141] How to chart a course of policy action between these two extreme-one is a politico-military construct of state security and the other is environmental security as national security ? "A unifying conceptual" framework "is needed from which to overview the complex relationships and tradeoffs among domains of life and values held by political communities such as nation states". Instead of giving a primacy to either environmental security, military or any other threats, "we should attempt to level the playing field in order to better appreciate the side effects associated with particular choices and how best to allocate scare collective resources." [142] Psychological and organizational constraints stand in the way of achieving such a unifying overview. Alexander George identified the perennial problems of value complexity in policymaking. Policy problems "are laden with competing values and interest" and "the standard textbook model of rationality cannot be employed in such instances, because the multiple values embedded in the policy problems cannot be reduced to a single utility function that can then be used as criterion for choosing among options".[143] It so happens in the decision making process that sometimes the core values go unnoticed only to be identified with hindsight. "Even where a range of competing values are identified by a decisionmaking process, there are strong organizational and cognitive pressures towards suppressing value conflict rather than confronting it".[144] A number of psychological theories and empirical research findings suggest that individuals betray a subconscious need for cognitive consistency.[145] Yugo Vertzberger has argued that "people see the social environment as consistent and balanced". As a result, 'they believe that a policy that serves one value is also likely to contribute to other values, thus avoiding consideration of value trade offs', and "even salient internal contradictions are often ignored".[146] In order to deal with these problems Ralph Kenney propounded a radically different decision paradigm: " value focused thinking". In this paradigm : significant effort is devoted to articulating values", prior to other activities. Then, " the articulated values are explicitly used to identify decision opportunities and to create alternatives".[147] In his view, such a value-focused orientation will help " create better decision situation with better alternatives which should lead to better consequences."[148] According to Stern, the concept of comprehensive security formulated by Barry Buzan, Arthur Westing, and others promises to serve this important integrating function by explicitly recognizing the multiple and interrelated dimensions and core value clusters of domestic and international life".[149] Buzan's formulation is useful because it includes military, economic, environmental, societal, and political dimensions. In the first edition of Barry Buzan's People, States and Fear, was evident that security at the three levels individual, state and international were central to the argument, but in some sense national security was still privileged. "Did Buzan want to make a three-decker out of the concept of security, or was it a conceptualization of national security" ?[150] It was clarified in the second edition (1991). The state level is privileged, and the argument about the levels is the absolutely correct one that national security cannot be comprehended at the state level alone : the dynamics of national security can only be grasped by seeing the interaction with processes at the societal / individual and the international level. Thus the argument is no longer one for "the right balance between the three levels", but one for enlightened national security taking in dynamics at all levels.[151] "The security label is one solution", says Barry Buzan – but tends himself towards recommending the other path : to 'identify by environmental issues as part of the economic agenda" which has the "advantage of sitting the issue at the heart of the action that is most relevant to it. There might, in the long run, be more advantage to making producers, consumers, tax men and economists factor environmental costs into their accounting activities, than to arming the state with emergency powers derived from an analogy with war. It might be argued that process type threats are better met by the process type remedies of economics, than by the statist solution of security logic."[152] In Buzan's analysis is found dominant the economic content approach to environmental crisis, though the security label at the state seems to be privileged. His later writings after the end of the cold war seem to have shifted the security label down from the state, but the state "remains central but no longer dominates either as the exclusive referent object or as the principal embodiment of threat in the way it did previously".[153] A spurt of a new range of referent objects for security and sources of security is found above, below and alongside the state. what ascend above the state to the status of referent objects are new international regimes, rules, norms and institutions on matters ranging from climate change, to economic liberation and proliferation of weapons of mass destruction. If the predictions of scientists about the imminence of climate change, major changes in temperature and precipitation pattern due to global warming become certain and accurate within a few years, Buzan does not mind keeping environmental security at the top of the global agenda. A dilution of his earlier stand that attaches the privileged security label at the state takes place. In the environmental sector the range of possible referent objects is very large, ranging from relatively concrete things such as the survival of individual species (animals, human kind) or types of habitat (rain, forest, lakes) to the maintenance of the planetary climate and biosphere. The relationship between the human species and the rest of the biosphere not only within the present time frame but also beyond timelessness and carried into the future remains central to whether that relationship can be sustained without risking either or both resulting in the collapse of the achieved levels of civilization or a wholesale disruption of the planet's biosphere. At either the macro or micro extremes" of interplay amongst all of these things "there are some pretty clear cases of existential threat (the survival of species, the survival of human civilization) that can be securitized."[154]

Critical Security Studies

It studies security in constructive terms. It challenges the traditionalists and wideners by applying post positivist perspective, such as critical theory and post structuralism.[155] Much of the work on this like Buzan's deals with social reconstruction of security. But the Critical Security Studies show that change is possible because things are socially constituted. If states dominate the arena, this is a feat of power politics repressing other dimensions of reality that would potentially replace the states if an emancipatory praxis with the help of critical theory empowers other subjectivities than those dominate at present. The social world does not exhibit any inexorable iron Jaws, all regularities can be broken, and it is the task of critical theory to show this. On this issue of security, the Critical Security Studies *is* often less constructive. "As part of the argumentation against established discourses of security, it will often try to mobilize other security problems—environmental problems, poverty and unemployment as more important, more threatening and thereby relativizing conventional wisdom. By this method they often end up reproducing the traditional and objectivist concept of security.[156] As observed by Waever, this approach will often contribute to the general securitization of ever large spheres of social life.[157] But the critical theory remains half way emancipatory without taking into the mainstream of its thinking ecological crisis or the environmental movement as sources of emancipatory change. They have not yet pursued the many insights offered by ecological thought.

Treating nature as a whole while observing its rich diversity, the critical theory would appreciate the patterns of mutual aid and complexity of natural phenomena, and question arguments based on the ontological stratification or linear thinking.[159] Holism would not associate with global control and homogenization but with an expanded self.[160] How the widening view of security proposed by critical theorists is possible for its emancipatory mission without the environment forming its mainstream thinking? In this chapter the above discussion of ecological critique of both liberalism and realism forming the mainstream of the dominant international perspective on security shows, the critical theorists need to make a review of the dominant hegemonic portrayal of human nature, its relationship with nature and a dismantling of 'self help versus the other syndrome' as a basis of redefining its security from a widening perspective. The ecological approach gets at the very root of emancipation by revealing the relationship between the control of nature and control of man/woman. In this sense, ecology is the holistic approach par excellence. By definition, it cannot limit its defence of freedom to specific constituencies, for no long term freedom may be gained for any group at the expense of nature; nature is nothing but the larger self of human kind.[161]

Conclusions

As the above discussion shows, despite cataclysmic changes in International Relations, environment and ecology, a dominant group remains adamant and unwilling to disrobe themselves of Westphalian straightjacket while understanding the diverse strands and security perspectives, the source of security threats no longer lie embedded in the state.

At the core of this .mainstream thinking and portrayal of International Relation remains dominant the ontology of an egoism-anarchy thematic The boundaries of what is ethically possible are established within a rigid universal closed power politics logic, which not surprisingly, gives politico-ethical legitimacy to great power dominance and hegemonic systems of global order. Thinking beyond this thematic of any unconventional seems to be anathema to this school of thought. The deepening environmental crisis with catastrophic consequences on society and human beings and their views of world hastened to the center stage to challenge the dominance of the statist school of thought customarily inclined to view security on politico-military terms. When the new threats from unconventional sources i.e from environmental degradation were put to fit into their Westphalian procrustean bed, and could not match their logic and parameters, these are termed as dependable variables and not causal and cannot ascend to the status of national security. Rather view them as predominant and separate security threats from the environmental crisis, they try downplaying these threats and associate these with political and economic factors as Buzan does. Another thing that is mostly ignored is the very assumption that when environmental security is held as national security, it does not mean that it seeks solution to the environmental crisis through the military apparatus of the state only. What is central to the realist logic is that the response to the threats if done through the exercise or use of the force is worth being named as having acquired the status of national security. Deudney, Dalby and others fearing their stand slipping into parochial statism have set all argumentations against bestowal of national security status to environmental threats. On the other hand Homer Dixon and others like Myers, Libiswezski have shown empirically that environmental degradation and scarcity of resources will produce social effects in terms of poverty, decline in agricultural productivity, displacement of people and weakening of political and social institutions. i.e., state which in turn will culminate in various violent intra or inter state conflict such as simple scarcity, group identify conflict, insurgencies and other forms of violence depending on the severity of the environmental crisis, the nature of the society and other contextual factors. their mapping out the scenario of violent conflict, from environmental scarcity of resources along its social effects, is not within the politico-military construct of national security. On the issues of environmental scarcity and degradation of renewable resources such as water, the conflict has taken the form of intra-state conflict such as insurgencies, group identity conflict having implications on international security and stability. The thrust of their argument rests on the fact that the environmental scarcity of resources unless properly addressed to has the dangers of violent inter and intra state conflict. Hence environmental degradation comes as a non-military security threat to the people of a state to occupy the status of national security.

National security understood in parochial and conventional wisdom needs to be redefined in view of the disastrous transformation and changes in environmental sector of the planet. The referent object of security is no longer the state, but other referent objects have appeared to be taken note of by the security strategists. These are the people, the individuals, the non human beings, future generations, the biotic community and the integrity of the cosmos. The trend towards dissociating security and nation state now clearly visible in world politics has been "strong enough to avoid the dangers of falling down the slippery slope toward a parochial statism that Deudney , Dalby and others fear".[162] The confinement of the debate over whether environment and conflict linkage assumes the status of national security to the Westphalian logic and wisdom cannot be a solution to the environmental crisis. If environmental degradation like the military can be assumed to be a threat and needs a military response, then confinement of the issues to the power politics of the sovereign states system, blurs the distinction between the two and aggravates the crisis. The hegemonic, homogenization and lineraity logic typical of world power structure and arrangements comes in to dominate the issue and not to find a solution with different, and alternative parameters that would challenge the above conventional logic and wisdom.

Going beyond the dominant environment and conflict paradigm, the very logic of the concept environmental security suggests a solution not in military but in non-military terms. This means traversing beyond the strict allegiance to the sovereign state system. This is what suggested by Stern :

"Blind allegiance to the nation state system and inability to contemplate shifts of sovereignty 'upward' or 'downward' needlessly limit the range of institutionalised or adhoc security strategies, which may be brought to bear in averting or coping with military and non military threats to individual and collective well being." [163]

Looking beyond the sovereign state system means assurance of environmental security to all. It is a kind of positive peace not based on violence, or just absence of interstate war but focuses on dismissal of hegemonic, ordering, fixity and certainty connotations of the present international system which is responsible for much of human fear, suffering and ecological disaster.[164] Ecological ethics and insight on which environmental security is based can offer purpose and direction to human kind by stressing natural cycles as well as the cooperative tendencies of nature.

Environmental security envisages a political project, and reconstruction of human society for human kind in biological imperatives, as aimed by critical theorist but not on ecological insights. It does not assume that human relation are fixed and therefore cannot accept the principle of cultural superiority. In a sense ecological insights "evoke the notion of natural law, but as a much different sort of essentialism divested of hierarchical pretence".[165]

The ecological insight challenges the ethnocentric western paradigm of development, its anti nature and perception of other as different from self, its inordinate life style, and exorbitant consumption and philosophy of life based on pure materialism, commodious living and production of wealth at the expense of others (nature, human beings, non-human living and non living, future generation). Pursuance of this hegemonic ideology and its universalization in the name of liberal economic emancipation, globalization and world order has been the root causes of the pauperization and despoliation of the planet. The proponents of environmental security and its critique and even the critical theorists have never touched upon this point and seem to have displayed the cultural superiority bias of hegemonic mainstream of International Relation. The author in this work stresses upon a new planetary wisdom and vision that will be notably the foundation of environmental security but also the national interests of each and every sovereign state. The imperatives of environmental security must bring to realization of policy makers, that 'emphasizing environmental values at a variety of levels including the national may help drive home the insight that self help nationally based strategies will probably not prove a sufficient basis for coping with the environmental threats now on the horizon".[166]

Environmental threats assume the status of national security on the basis of what Homer Dixon and others propounded, and what Deudney denounced. Dixon's mapping out of inter and intra state conflicts arising out of environmental scarcity and degradation of resources assume the status of national security in Westphalian logic with the emergence of new referent objects of security other than the sovereign state, the environmental degradation affecting these referents will be perceived as security threats demanding not solution in terms military frame work of sovereign state system. The very survival of the people, eco system and bio regions demands new planetary ethics, which ultimately will define the national security interests. The redefinition of national security prompted by environmental degradation replacing the traditional understanding of national security makes the Deudney's disparagement of Homer Dixon and other's analysis discredited, if protection of the people against the threats stemming from non-conventional sources, i.e., the environmental degradation –the consequences of which are more disastrous than the war itself.

What emerges significant is to construct a way rethinking the debate around the environment and sovereignty, which moves beyond what

Karen Litfin, calls the geological model of sovereignty being eroded by environmental change and environmental regimes.[167] The image of sovereignty in Litfin's work, The Greening of Sovereignty in World Politics, as a fixed thing, akin to a rock being clearly recognized as socially constructed and consisting of a bundle of rights whose composition and location shift continuously across time. Further the study suggests that sovereignty should be conceived in terms of three elements- autonomy, control and authority- all of which can be attached to different institutions overtime not necessarily just states, and be constructed themselves in differing ways.[168]

Moving beyond the western straitjacket , the concept 'environmental security' envisions :

- Rethinking western pattern of development, its cultural superiority, hegemonic power structure and universalization doctrine.
- Replacement of the mechanical , the other view of nature and human beings by an organicism view which studies human beings, nature, future generations and all interconnected, coordinated in a symbiotic relationship as members of a cosmos.
- Global governance of environmental matters through a coordinated network of NGOs, environmental regimes, norms, regional organizations, and communities cutting across the rigidity of sovereign state system. This means not a proselytization but shifting of sovereign state system to other centers, at every level. The traditional unidimensional view of sovereignty appears superficially to be inconsistent with the multidimensional requirements of planetary environmental protection.
- Pronouncement of cultural diversities, differentiated civilizational values and not deification of western values.
- An ethical and moral rebuilding of human nature on a spiritual plank to think in terms of an expanded self within a timelessness frame.

References

1. Kenneth W. Waltz, Theory of International Politics (Reading, Massachusetts: Addison-Wesley, 1979), p-102.

2. Richard N. Lebow, "The Long Peace, the End of the Cold War, and the Failure of Realism", International Organization, Vol. 48, no. 2, 1994, pp. 249-77.

3. Michel Frederick, "A Realist's Conceptual Definition of Environmental Security", in Daniel H. Deudney and Richard A. Matthew ed., Contested Grounds: Security and Conflict in the New Environmental Politics (Albany : State University of New York Press, 1999), p. 92.

4. Charles Philippe David, "Strategic Studies in Crisis". Etudes internationales, Vol. 20, No. 3, 1989, pp. 513-15.

5. Barry Buzan, "Change and Insecurity : A Critique of Strategic Studies", in Barry Buzan and R.J Barry Jones, eds. Change and the Study of the International Relations : The Evaded Dimension(New York : St. Martin's Press, 1981), Lester Brown, "An Untraditional View of National Security", in John Reichart and Steven Sturm, eds, American Defence Policy, 5th edn (Baltimore, Md. : Johns Hopkins University Press, 1984).

6. Norman Myers, "Environmental Security" Foreign Policy, Vol. 74, 1989, pp. 23-41.

7. Richard Ullman, "Redefining Security", International Security, Vol. 8, no.1, 1983, pp. 129-53

8. Michel Frederick, *op.cit.*, p. 94

9. Michel Frederick, *op.cit.*, p. 96

10. Sverre Lodgaard, "The Transformation of Europe and the Gulf War : Implications for the Research Agenda". UNIDIR Newsletter, Vol. 4. no. l. 1991, pp.7-9.

11. Arthur H.Westing, "Comprehensive Human Security and Ecological Realities" Guest Comment. Environmental Conservation, Vol. 116, Winter 1989, p. 295.

12. Daniel B.Botkin, Discordant Harmonies : A New Ecology for the Twenty First Century (New York : Oxford University Press, 1990); Rice Odell, Environmental Awakening : The New Relation to Protect the Earth (Cambridge, Mass.: Ballinger, 1980).

13. Johan J. Holst, " Security and the Environment" : A Preliminary Exploration, Bulletin of Peace proposals Vol. 20, no. 2, 1989, pp.123-28.

14. *Ibid.*, p.123.

15. M. Frederick, *op.cit.*, p.98.

16. Barry Buzan, "Environment as a Security Issue", Geopolitical Perspectives on Environmental Security, Paul Painchaud, ed. (Quebec: Laval University, GERPE, Occasional Paper, 92-05,1992), p. 4

17. Raymond Aron, Peace and War : A Theory of International Politics (New York : Doubleday, 1966), pp.72f and 598f.

18. Ole Waever, Securitization and Desecuritization, Working Papers, Centre for Peace and Conflict Research, Copenhagen 5,1993, p. 5.

19. Barry Buzan , "Environment as a Security Issue", *op.cit.*, p. 4.

20. Ole Waever, Securitization and Desecruritization, *op.cit.*, p.5

21. Barry Buzan, People, States and Fear : An Agenda for Security Studies in the Post-Cold War Era (Hemel Hempstead : Wheatsheaf, 1991), p.117.

22. Egbert Jahn, Pierre Lemaitre and Ole Waever, European Security: Problems of Research on Non-Military Aspects (Copenhagen: Copenhagen Papers of the Centre for Peace and Conflict Research, 1987) p. 9.

23. Michel Frederick, *op.cit.*, p.100.

24. Prue Taylor, "The Failure of International Environmental Law", Our Planet, Vol.4, no.3, 1992,pp.14-15.

25. Arthur H. Westing, ed; Comprehensive Security for the Baltic: An Environmental Approach (London: Sage, 1989), pp.1-13.

26. L.H. Brown, "Regional Collaboration in resolving third world conflicts", Survival, (London) Vol. 28, 1986, pp. 208-220. See also L.J. Cantori and S.L. Spiegel, ed., International politics of Regions: A Comparative Approach (New Jersey: Prentice Hall, 1970), p. 432; B.M. Russett, International Regions and the International System: A Study in Political Ecology (Chicago: Rand –McNally, 1967), p.252.

27. Michel Frederick, *op.cit.*, p.101.

28. Daniel Deudney, "The Case against Linking Environmental Degradation and National Security", Daniel Deudney, "Environmental Security: Muddled Thinking", Bulletin of the Atomic Scientists Vol.47, no.3, 1991, pp.22-28.

29. Clive Ponting, A Green History of the World : The Environment and the Collapse of Great Civilizations (New York : St. Martin's, 1992).

30. SIPRI, Warfare in a Fragile World: Military Impact on the Human Environment (London: Taylor and Francis, 1980).

31. T.W. Hawley, Against the Fires of Hell: The Environmental Disaster of the Gulf War (New York: Harcourt Brace Jovanovich, 1992). Arthur H. Westing, " Environmental and Ecological Consequence of War", Sent on email to author on 15th August 2003, p. 4.

32. Paul R. Ehrlich, Carl Sagan, Donald Kennedy, and Walter Orr Roberts. The Cold and the Dark : The World After Nuclear War (New York : Norton, 1984).

33. Animesh Roul, Environmental Effects of the Production and Use of Weapons of Mass Destruction, Dissertation submitted to Jawaharlal Nehru University, for M.Phil, New Delhi, 1998. See also Michael Renner, "Assessing the Military's War on the Environment", in Lester Brown, ed., State of the World 1991 (New York: Norton, 1991), pp.132-52.

34. Daniel H. Deudney "Environmental Security: A Critique", in Daniel H.Deudney and Richard A. Matthew ed., Contested Grounds (Albany: State University of New York Press, 1991), pp.192-93.

35. Deudney, *Ibid.*, p.193

36. Thomas F. Homer Dixon, Environment, Scarcity and Violence (New Jersey/ Princeton: Princeton University Press, 1999), pp.142-47.

37. Deudney, *op.cit.*, p.194.

38. Deudney, *op.cit.*, p.194.

39. Deudney, *op.cit.*, p.194.

40. William James, "The Moral Equivalent of War", Memories and Studies (New York and London : Longmans, Green & Co., 1911).

41. Ken Conca, " In the Name of Sustainability: Peace Studies and Environmental Discourse", in Jyrki Kakonen, ed., Green Security or Militarized Environment (Aldershot, U.K. : Dartmouth, 1994).

42. Kofi Annan, We the Peoples: the Role of the United Nations in the 21st Century, March 2000, quoted by Andrew Mack "Human Security in the New Millennium", Work in Progress: A Review of Research Activities of the United Nations University Vol. 16, no. 3, Summer, 2002, p. 4

43. Jyrki Kakonen,ed., Green Security or Militarized Environment, *op.cit.*

44. Michael Howard, "War and the Nation-State", Daedalus, Vol. 108, no. 4, Fall 1979, pp. 101-10.

45. Charles Tilly, "War Making and State making as Organized Crime" in Peter Evans, Dietrich Rueschemeyer and Theda Skocpal, eds., Bringing the State Back in (Cambridge, UK. : Cambridge University Press, 1985), pp.169-91.

46. Jason Clay, "Resource Wars: Nation and State Conflicts of the Twentieth Century" in Barbara Rose Johnston ed, Who Pays the Prize? The Sociocultural Context of Environmental Crisis (Washington D.C. : Island Press, 1994).

47. Nancy Peluso, Rich Forests, Poor People (Berkeley : University of California Press, 1993), and Charles S.Wood and Marianne Schmink, " The Military and the Environment in the Brazilian Amazon", Journal of Political and Military Sociology, vol. 21, Summer 1993,pp. 81-105.

48. See Thomas F. Homer Dixon, Environment, Scarcity and Violence, *op.cit.* p.144.

49. Anil Agarwal, Sunita Narain, Anju Sharma and Achila Imchen, eds., Poles Apart : Global Environmental Negotiation 2, (New Delhi : Centre for Science and Environment, 2001).

50. Michael Ross Fowler and Julie Marie Bunck, Law, Power and the Sovereign State : The Evolution and Application of the Concept of Sovereignty (University Park : Pennsylvania State University Press, 1995).

51. Barry Buzan, "Rethinking Security after the Cold War", Nordic Journal of International Studies, Vol. 32, no. 1, March 1997, p. 12.

52. On environmental conditionality, See Andrew Hurrell 'Green Conditionality', Overseas Development Council Policy Paper, March 1993.

53. See Patricia Birnie, "International Environmental Law: Its Adequacy for Present and Future Needs", in Hurrell and Kingsbury eds., The International Politics of the Environment (Oxford : UK : Clarendon Press, 1992), p. 84.

54. Kathryn Sikkink, "Human Rights, Principled Issue- networks, and Sovereignty in Latin America", International Organization, Vol. 47, no. 3, Summer 1993, pp. 411-41, and Ronnie D. Lipschutz, "Reconstructing World Politics : The Emergence of Global Civil Society", Millennium Journal of International Studies, Vol. 21, no.3,Winter 1992,pp.389-420.

55. See Hurrell and Kingsbury, "introduction" in Hurrell and Kingsbury, eds., *op.cit.*, pp. 6-8.

56. Mark A Levy, Robert O keohane, and Peter M.Haas, "Imposing the Effectiveness of International Environmental Institutions" in Haas, Keohane and Levy,eds., Institutions for the Earth : Sources of Effective International

Environmental Protection (Cambridge, MA : MIT Press,1993), pp. 415-17.

57. Ken Conca, "Rethinking the Ecology Sovereignty Debate", Millennium: Journal of International Studies Vol. 23, no. 3, 1994, p. 702.

58. Ken Conca, *op.cit.*, p. 703.

59. Quoted from Ruth Lapidoth, "Sovereignty in Transition", Journal of International Affairs Vol. 45, no. 2, winter 1992, pp. 325-46.

60. Conca, *op.cit.*, p. 706

61. Deudney, *op.cit.*, p. 198.

62. Summit "Documents Safeguard Brazilian Interests", Daily Report : Latin America, FBIS-LAT-92-114-5, 12 June 1992, p. 27.

63. David Cleary , "After the Frontier : Problems with Political Economy in the Modern Brazilian Amazon", Journal of Latin American Studies Vol. 25, part. 2, May 1993, pp. 331-49.

64. Conca, *op.cit.*, p.711.

65. *Ibid.*, p. 311.

66. Quoted Kennedy Graham, "Leadership ethics for the future : Trend lightly on the Planet", Point of view, UNU nexions, April 2002, p. 5.

67. Matthew Paterson, "Interpreting trends in global governance", International Affairs, Vol. 25, No. 4, 1949, p.795; and Lamont Hempel, Environmental governance; the global challenger (Washington D.C: Island Press, 1996).

68. Murray Bookchin, The Ecology of Freedom : The Emergence and Dissolution of Hierarchly, Revised Edition, (Montreal and New York, Ny : Black Rose Books, 1991).

69. Joshua Karliner, The Corporate Planet : ecology and politics in the age of globalization, (San Francisco : Sierra Club Books,1997), Chapter-7.

70. Matthew Paterson, Understanding global environmental politics: domination, accumulation, resistance (London : Macmillan, 2000).

71. Lynton Keith Caldwell, International Environmental Policy : Emergence and Dimension (New Delhi : Affiliated East-West Press pvt. Ltd., 1991).p. 68.

72. Deudney, *op.cit.*, p.199.

73. See William Thomas, Scorched Earth: The Military's Assault on the Environment (Philadelphia, PA. and Gabriola Island: New Society Publishers, 1995). See also Matthias Finger. " The Military. the Nation State and the Environment", The Ecologist. Vol. 21. no. 5. 1991. pp. 220-25.

74. John Barkdull, "Why Environmental Ethics Matters to International Relations", Current History Vol. 99. No. 640. November-2000. p. 362.

75. John A.Vasquez, "The Post-Positivist Debate : Reconstructing Scientific Enquiry and International Relations Theory After Enlightenment's Fall", in Ken Booth and Steve Smith ed., International Relations Theory Today (Cambridge, Polity Press, 1995), p. 221.

76. Eric Laferriere , *op.cit.*, p. 68.

77. Kenneth N. Waltz, Man, the State and War : A Theoretical Analysis (New York, NY : Columbia University Press, 1959).

78. Eric Laferriere, *op.cit.*, p. 68.

79. *Ibid.*

80. Barry Buzan, " Rethinking Security after the Cold War", Nordic Journal of International Studies Vol. 32, No.1, March-1997. p. 19.

81. J.R. McNeill, "Ideas Matter : A Political History of the Twentieth- Century Environment", Current History Vol. 99. No. 640. November. 2000, p. 374.

82. Richard Cobden, Political Writings, Vol. I (London : T Fisher Unwin,1903). Norman Angell, The Great Illusion(New York, NY : Knickerbocker Press. 1911).

83. David Mitrany, A Working Peace System (Chicago. IL : Quadrangle Books. 1966), Ernst B.Haas, Beyond the Nation-State : Functionalism and International Organization (Stanford, CA : Stanford University Press, 1969).

84. Robert O.Keohane, After Hegemony : Cooperation and Discord in the World Political Economy (Princeton, NJ ; Princeton University Press, 1984) and Stephen D Krasner, ed., International Regimes (Ithaca, NY:Cornell University Press,1983).

85. Robert C.Johansen, The National Interest and the Human Interest : An Analysis of US Foreign Policy (Princeton . NJ : Princeton University Press. 1980).

86. Edward Goldsmith, "Development as Colonialism". World Affairs. Vol. 6. no. 2. April-June 2002. p. 36.

87. Ismail Shariff and David M.Littig, "Globalization old wine into new bottles", World Affairs, Vol. 6, No. 2, April-June 2002, p. 42.

88. See Karl Deutsch et al., Political Community and the North Atlantic Area (Princeton, NJ ; Princeton University Press, 1957), and Amitai Etzioni, Political Unification : A Comparative Study of Leaders and Forces (New York, NY : Holt, Rinehart and Winston, 1965).

89. James N. Rosenau, Turbulence in World Politics : A Theory of Change and Continuity (Princeton, NJ : Princeton University Press, 1990), p. 289.

90. Mitrany, *op.cit.*, p. 96, Hoarold Laski, A Grammar of Politics (London : Allen and Unwin, 1967), p. 614; See also David Long, "International Functionalism and the Politics of Forgetting", International Journal, Vol. 48. no.2, 1993, pp. 356-79.

91. Bruce Russett, "Causes of Peace" in Carolyn M Stephenson ed., Alternative Methods of International Security (Washington, D.C. : University Press of America, 1982), pp. 188 & 191.

92. Mitrany, *op.cit.*, p. 36.

93. Eric Laferriere, *op.cit.*, p. 72.

94. Origins : NC Documentary Service 17, 3 March 1988, p. 645 ; Quoted in J. Ronald Engel, "Introduction : The ethics of sustainable development", in J.Ronald Engel and Joan Gibb Engel ed., Ethics of Environment and Development (London : Belhaven Press, 1990), p. 4.

95. Stephen Toulmin, The Return to Cosmology : Postmodern Science and the Theology of Nature (Berkeley : University of California Press, 1982), pp. 255-56.

96. Holmes Rolston, III, Environmental Ethics : Duties to and Values in the Natural World (Philadelphia : Temple University Press, 1988), p. 174.

97. Carolyn Merchant, The Death of Nature : Women, Ecology and the Scientific Revolution (New York, NY and Toronto : Harper and Row, 1980).

98. Janet Biehl, Finding Our Way : Rethinking Ecofeminist Politics (Montreal and New York, NY : Black Rose Books, 1991), pp. 89-90.

99. Barry Buzan and Richard Little, "Why International Relations has Failed as an Intellectual Project and What to do about it", Millennium; Vol. 30, no. 1, 2001, pp. 25.

100. *Ibid.*, p. 26.

101. Deudney, "The Case against Linking Environmental degradation and National Security", Millennium vol 1990, pp. 461-76.

102. Eric K.Stern, " The Case for Comprehensive Security", in Daniel H.Deudney and Richard A. Matthew, eds., Contested Grounds, *op.cit.*, p. 139.

103. Barry Buzan, Peoples, States and Fear : The National Security Problem in International Relations (Chapel Hill : University of North Carolina Press, 1983), p. 245.

104. Eric K.Stern, *op.cit.*, p. 139.

105. *Ibid.*

106. Ken Booth,"Security in Anarchy: Utopian Realism in Theory and Practice." International Affairs Vol. 67, no. 3, 1991, pp. 527-45.

107. *Ibid*, p. 140.

108. Stephen Walt, "The Renaissance of Security Studies", International Studies Quarterly Vol. 35, no. 2, 1991, p. 212.

109. Joseph Nye and Sean Lynn Jones. "International Security Studies: A Report of a Conference on the State of the Field," International Security vol. 12, no. 4, 1988, p. 6.

110. Walt, "The Renaissance of Security Studies", p. 213.

111. Stern, *op.cit.*, p. 135

112. Walt, *op.cit.*, p. 135

113. Stern, *op.cit.*, p. 135

114. See M D'Antonio. Atomic Harvest: Hanford and the Lethal Toll of America's Nuclear Arsenal (New York: Crown Publishers, 1993); and Murray Fesbach. Ecological Disaster: Cleaning up the Legacy of the Soviet Regime (New York; Twentieth Century Press, 1995).

115. See David Baldwin, Economic Statecraft (Princeton; NJ: Princeton University Press, 1985).

116. T.W Hawley, Against the Fires of Hell: the Environmental Disaster of the Gulf War (New York : Harcourt Brace Javonovich, 1992), p. 8.

117. Stern, *op.cit.*, p. 136

118. N.Brown, "Climate, Ecology, and International Security", Survival Vol. 31, no. 6, 1989, pp. 519-32.

119. S. Lodgaard, "Environment; Confidence Building and Security", in Hjort af Ornas and Lodgaard, eds., The Environment and International Security (Uppsala: PRIO/Uppsala University Program on Environment and International Security, 1992), pp.11-22.

120. Barry Buzan, Peoples, States and Fear, 2nd edt, *op.cit.*

121. Stern, *op.cit.*, p.138.

122. *Ibid.*

123. Marc A Levy, "Is the Environment a National Security Issue"? International Security Vol. 20, no. 2, 1995, p. 39.

124. Eric K.Stern, "High Politics and Crisis: Military and Environmental Dimensions of Security", in Hjort af Ornas and Laggard ,eds. ,The Environment and International Security, *op.cit.*, p. 90

125. Levy, "Is the Environment a national Security issue"? p. 60.

126. *Ibid.*, p. 61.

127. See for details, Eric K.Stern, "The Case for Comprehensive security", *op.cit.*, p. 141.

128. Levy, "Is the Environment a National Security Issue"?, *op.cit.*, p. 50.

129. *Ibid.*

130. Stern, "The Case for Comprehensive Security", *op.cit.*, p. 142

131. Levy, "Is the Environment a National Security Issue?" *op.cit.*, p. 54.

132. Ibid, p. 36.

133. Stern, "The Case for Comprehensive Security", *op.cit.*, p.142.

134. Stephen M. Walt, "The Renaissance of Security Studies", International Studies Quarterly Vol. 35, no. 2, 1991, pp. 212-13.

135. Barry Buzan, Environment as a Security Issue" in Paul Painchaud ed. Geopolitical Perspectives on Environmental Security, Cahier du GERPE, No.92-05,Universite Laval,Quebec, pp. 1 and 24f.

136. Barry Buzan, "Rethinking Security after the Cold War", Nordic Journal of International Studies Vol. 32, no. 1, March 1997, p. 11.

137. Barry Buzan, n. 135, p. 26.

138. Daniel Deudney, "The Case against Linking Environment Degradation and National Security" *op.cit.*, p. 469.

139. Richard H.Moss, "Environmental Security ? The illogic of centralized state responses to environment threats", in Paul Painchaud, ed., Geopolitical Perspectives on Environment Security, *op.cit.*, p. 24.

140. Moss, *op.cit.*, p. 32.

141. Patricia M. Mische, "Ecological Security and the Need to Reconceptualize Sovereignty", Alternatives Vol.14, 1989, pp. 389-427.

142. Stern, "The Case for Comprehensive Security", *op.cit.*, p. 143.

143. Alexander L.George,Bridging the Gap : Theory and Practice in Foreign Policy (Washington, D.C. : US Institute of Peace Press, 1993), p. 27.

144. Stern, "The Case for Comprehensive Security", *op.cit.*, p. 143.

145. Irving Janis and L.Mann, Decision making : A Psychological Analysis of Conflict, Choice, and Commitment (New York: Free Press.1977).

146. Y.Vertzberger, The World in Their Minds (Palo Alto, California : Stanford University Press,1991), pp. 138-39.

147. R. Kenney,Value-Focused Thinking : A Path to Creative Decision making (Cambridge, Mass : Harvard University Press,1992).

148. *Ibid.*, p. IX

149. Stern, "The Case for Comprehensive Security", *op.cit.*, p.144.

150. Ole Waever, Securitization and Desecuritization, *op.cit.*, p.3.

151. Egbert Jahn, Pierre Lemaitre and Ole Waever, European Security : Problems of Research on Non-Military Aspects, *op.cit.*, pp.51-53.

152. Buzan , "Environment as a Security issue" in Paul Painchaud ed., Geopolitical Perspectives on Environmental Security, *op.cit.*, p.25.

153. Barry Buzan, "Rethinking Security after Cold War", *op.cit.*, p.11.

154. Barry Buzan ,Rethinking Security after the Cold War", *op.cit.*, p.18.

155. Keith Krause and Michael C.Williams, "From Strategy to Security : Foundation of Critical Security Studies" in Krause and Williams , Critical Security Studies (Boulder, CO : Lynne Rienner,1996).

156. Barry Buzan, "Rethinking Security ..." *op.cit.*, p.19.

157. Ole Waever, "Securitization and Desecuritization" in Ronnie Lipschutz ed., On Security (New York : Columbia University Press, 1995), pp.46-86.

158. Eric Laferriere, "Emancipatory International Relations Theory : An Ecological Perspective", Millennium : Journal of International Studies Vol. 25, no. 1, 1996.

159. Peter Kropotkin, Mutual Aid : A Factor of Evolution (Boston, M.A : Extending Horizon Books , 1955).

160. Warwick Fox, Toward a Transpersonal Ecology : Developing New Foundations for Environmentalism (Boston , MA : Shambhala , 1990).

161. Eric Laferriere, *op.cit.*, p. 74.

162. Eric K.Stern, "The Case for Comprehensive Security", *op.cit.*, p. 145.

163. *Ibid.*

164. On Peace Research and International Relation see Beverly Neufeld, "The Marginalization of Peace Research in International Relations". Millennium Vol. 22. no. 2, 1993, pp. 165-84.

165. Eric Laferriere, *op.cit.*, p.74.

166. Eric K.Stern, "The Case for Comprehensive Security", *op.cit.*, p.145.

167. Karen Litfin, "The greening of sovereignty : an introduction", in Litfin, ed., The Greening of Sovereignty in World Politics (Cambridge , MA : MIT Press, 1998).

168. *Ibid*, See also Karen Litfin, "Sovereignty in World Politics", Merskon International Studies Review 41, 1997, pp. 167-204.

4

Environmental Scarcity of Resources

Starting from 1st world war not only to the Falkland war as enumerated by Westing, but to the latest Gulf War all wars have been fought over access to natural resources such as minerals, fuels, fish stocks, products of the land and the land itself. The whole history of humanity has borne a testimony to the congeries of violent conflicts, contentions and wars that had taken place over access to and distribution of natural resources. In view of this, many scholars exhibit a tendency to interpret the linkage between environment and conflicts as a question of struggle for scare natural resoures.[1] Though environmental conflicts can manifest themselves as conflict over resources, in fact these multitude of wars are not really coming within what the term 'environmental' is understood. Human ecology poses questions about the nature of the relationship between the humans and the environment. The basic problems of this relationship under the study of social ecology are the limits of the utilization of the natural resources and of pollution of air and landscapes. This also implies a number of limits in the demands made of and burdens laid on the human body and on the social life of the community.[2] A human made environmental change does not simply mean any interaction between humans beings and their environment. It means a destabilizing interference in the ecosystem's equilibrium. The pollution of atmosphere by emission of carbon dioxide, for example from fossil fuels based industries, overtaxes the ecosystem's absorption capability and causes the so called green house effects. This climate warming, in turn changes among other things, the balance between ice sheets and liquid water on the earth's surface, causing a rise of the sea level. In a similar way, the process of desertification can be explained as consequences of the devegetation and overuse of the soil causes a decline of precipitation and soil erosion. As water supply is the prerequisite of nearly every ecosystem's functioning, serious manipulation in river flows are another example of human interference which often affects the ecological equilibrium in the concerned regions. The environmental degradation is only about these affected life supporting conditions, the negative impact on human beings and society. Environmental change can mean a loss of quality, a degradation.[3] Environmental degradation is thus, understood as a human made environmental change having a negative impact on human society. Gunther Bachler, in the place of degradation used the word 'transformation'. The term 'global human – ecological transformation' is meant the introduction of a heuristic concept in regarding those interactions between the three levels- nature, human beings and the economy in the regional context-which have led to conflicts between human communities and will increasingly lead to such conflicts."[4]

To know the critical differentiation between renewable and non renewables- surprisingly not highlighted by many of the participants in this long debate- is necessary to understand the concept of environmental change and degradation. Natural resources can be roughly divided into two groups : non-renewables, like fresh water, soil, forests, air, climate, oceans and biodiversity. Renewable resources can be further divided into two groups : those that provide goods and those that provide services. Ocean fisheries supply fish that are a good while the stratospheric ozone layer renders a service by protecting life from high levels of ultraviolet radiation. Some renewables provide both goods and services ; forests supply timber (a good) while also maintaining regional hydrological cycles (a service).They are renewable because they are ecologically integrated in a feedback circle system which guarantees their replacement or the preservation of their quality. On the other hand non-renewable resources as they are, cannot be integrated in such an ecosystem.[5] Therefore they can be depleted but they cannot be degraded. Their depletion causes degradation of the renewable resources. Further to distinguish, a non-renewable consists of a stock, which is the total quality of the resource available for consumption. A renewable resource has both a stock and flow which is the incremental addition to or restoration of the stock per unit of time.[6] To draw an analogy between the stocks and flows of renewable resources and money in bank, a renewable resources's stock is the resource's capital or principal, which generates a flow of interests or income that can be tapped for human consumption and well-beings.[7]

A further characteristic of renewable resources is that they can be less substituted than non-renewable ones, and in several cases not substituted at all. Food, fresh water, clean air and life supporting climatic conditions are not just sources of wealth in an economic sense but also biological prerequisites of life. In addition to environment as a bank of material resources and a sink for wastes a third dimension of environment as a space of living should be considered.[8] The loss of settlement areas due to the rise of the sea level, to desertification, or to industrial accident or project or dams, means more than just a shortage of resource 'land'. It includes an existential dimension and in certain dramatic cases of environmental degradation, it implies an aesthetic dimension of environment's impoverishment.[9]

The concept of environmental degradation in order to be understood correctly, leads to another important differentiation concerning the concept of resource scarcity that are of four distinct types.[10]

1. Physical scarcity means that resource is available in a finite amount.
2. Geopolitical scarcity means that resources are often distributed unequally on the surface of the earth so that some countries depend on deliveries from others.
3. Socio-economic scarcity concerns the unequal distribution of purchasing power and of property rights to provide natural resources between or within societies.
4. Environmental scarcity caused by environmental degradation means that resources that have traditionally been regarded as plentiful and naturally renewable but are becoming scarce now because of this failure of human beings to adopt sustainable methods of their management.

Unequal geopolitical and socio-economic distribution are often a source of degrading behaviour and the physical scarcity of a renewable resource can be a reason for the depletion of resource's capital stock. These are causally interrelated. Here the main focus is on the environmental scarcity of renewable resources that has the potential to seriously undermine human well- being. It can arise in three ways : through a drop in the supply of a key resource, through an increase in demands, and through a change in the relative access of different groups to the resource. Homer Dixon calls these respectively supply induced, demand induced and structural induced scarcities. Depletion and degradation produce a decrease in total resource supply or in other words a decrease in the size of the total resource stock. But population growth and changes in consumption behavior can also cause greater scarcity by boosting the demand for a resource. Thus, if a rapidly growing population depends on a fixed amount of cropland, the amount of cropland falls inexorably. Finally, scarcity is often caused by a severe imbalance in the distribution of wealth and power that results in some groups in a society getting disproportionately large slices of the total resources, where as others get slice that are too small to sustain their livelihoods. This is what is called structural scarcity.

Analysts and policymakers in the past have usually addressed these three sources of scarcity independently. Neo Malthusians and economic optimistic have given importance to supply and demand induced scarcity while generally ignoring the political economy of resource distribution. According to Dixon, these three scarcities interact and reinforce each other in an extraordinarily pernicious ways.[11] One type of interaction is resource capture. The degradation and depletion of renewable resources can interact with population growth to encourage powerful groups within a society to shift resource distribution in their favour by manipulating laws and institutions governing resource access. These powerful groups tighten their grip on the dwindling renewable resources while subjecting the marginal groups to greater scarcity.[12]

Thus in Bangladesh, worsening land scarcities, partly caused by rapid population growth and human induced natural disasters, encouraged powerful landowners to exploit weakness in the state's land inheritance laws in order to seize lands from the poor farmers. These peasants have been forced to migrate to Indian territories for better livelihood. This happens also in Chiapas where the peasants were forced deeper into the state's lowland rain forest, farther away from the state's economic heartland, and deeper into poverty.

Thus in Bangladesh, warming land scarcities partly caused by rapid population growth and human induced natural diasters, encouraged powerful landowners to exploit weakness in the state's land inheritances laws in order to seize lands from the poor farmers. These peasants have been forced to migrate to Indian territories for better livelihood. This happens also in Chiapas where the peasants were forced deeper into the state's lowland rain forest, farther away from the state's economic heartland, and deeper into poverty.

Resource Capture

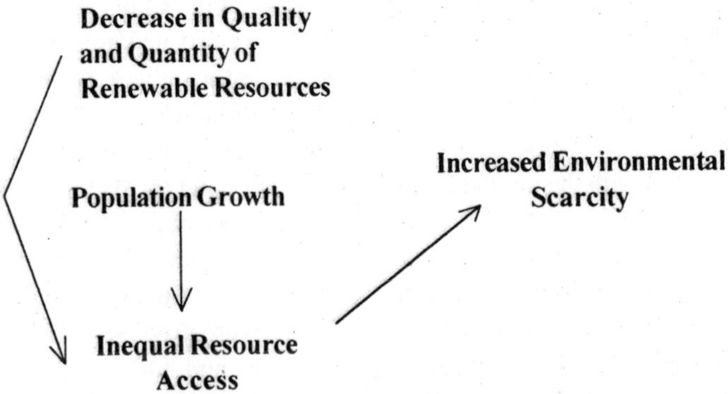

This figure is a representation of the process of resource capture as outlined by Homer Dixon.[13]

In the early 1990s, Israel's average annual supply of renewable fresh water was about 1950 million cubic meters against its demand including that of settlements in the West Bank and Golan Heights which was 2,150 cubic meters.[14] The annual deficit of about 200 cubic meters is covered by overpumping aquifers which has already plummeted water tables in some parts of Israel and West Bank. This decline has caused the exhaustion of some wells and the infiltration of sea water from the Mediterranean.[15] At the same time Israel's population is expected to increase from the present 6.0 million to around 7.5 million in the 2020.Accordingly, the country's water demand by 2020 is estimated at 2,800 cubic meters.[16]

Ground water constitutes sixty percent or about 1200 cubic meters of annual sustainable supply of water. The rest comes from the river flow, flood water, and waste water cycling. Two of the three main aquifers on which Israel depends lie principally underneath the West Bank. About 475 cubic meters or about 40 percent of the ground water Israel uses originates in this territory. To protect its water interests Israel strictly restricted water use by Jewish settlers and Arabs on the West Bank. On a per capita basis in the early 1990s, settlers consumed four times as much as Arabs. Of the 650 cubic meters only 125 cubic meters were allowed to be used by the

Arabs. Israel also restricted the number of wells Arab could drill in the territory, the amount of water Arabs were allowed to pump, and the times at which they could draw irrigation water. During occupation after 1967 Arabs were not permitted to drill new wells for agricultural purpose, although the Mekorot, Israel water company drilled more than thirty wells for settler's irrigation. The area of irrigated Arab farmland on the West Bank dwindled from 27 percent of the total cultivated area to 3.5 to 6 percent.[17] Arab agriculture in the regions got a setback because some Arab wells became dry or saline as a result of Israeli wells being drilled deeper nearby. These Israel water policies restricting water use to Arabs combined with the confiscation of agricultural land for settlers as well as Israeli restriction on Palestine agriculture propelled many West Bank Arabs to abandon farming and flee to towns to find either unemployed or become day labourers within Israel. No doubt many political, economic and ideological factors, operate to explain the present conflagration in this region but the links between these processes and unrest in West Bank appears reasonable to conclude that water scarcity and its economic effects contribute to the present crisis in the region.[18] Rising frustration in Palestinian community due to this factor appears to lie underneath the long ongoing Israeli Palestine territorial war.[19]

Ecological Marginalisation

It is a process resulted out of a structural imbalance in resource distribution which in association with rapid population growth forces resource starved people into ecological marginal areas, such as upland hillsides.[20]

Ecological Marginalization

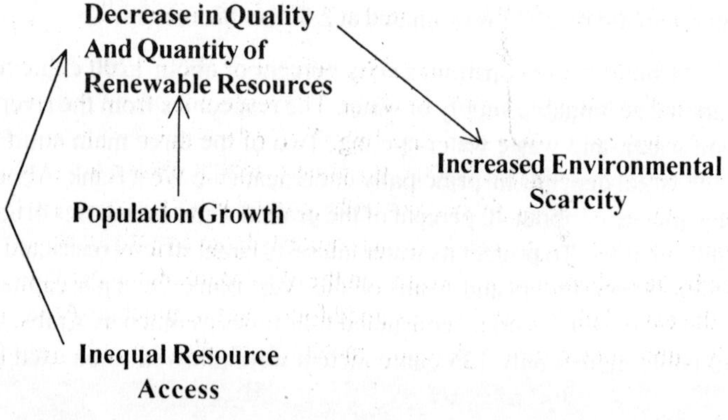

This figure as pointed out by Dixon represents the process by which poor people due to inequitable resource distribution, population growth and land degradation are forced to migrate to marginal regions.[21] As happened in Bangladesh, the people living in low lying areas affected by inequitable distribution of land, high density of population growth, soil erosion, and natural disasters such as cyclone, flood and storms have been settled in Chittagong Hill Tracts (CHT) which in turn, has not only caused massive deforestation because of slash and burn agriculture but also driven the indigenous tribal people Chakmas into Indian territories.[22] This process is visibly seen in the Himalayas, Indonesia, Central America, Brazil, Rajastan and the Sahel.Overtime the process " causes environmental degradation and intractable poverty to become more and more closely intertwined in particular geographic areas with fragile environmental degradation. The World's poorest people are thus increasingly clustered in two types of areas : remote and ecologically fragile rural areas and the edge of growing urban areas."[23] A grossly unfair distribution of good cropland in lowland regions a colonial heritance bequeathed to Philippines by its former Spanish and American colonial rulers continues to be reinforced since independence by a powerful landowning elite[24] Introduction of green - revolution and modern technologies in agriculture from 1960s into the 1980s resulted in a hike not only in the lowland production of grain for domestic consumption but also in cash crops. The proceeds from the cash crops helped the country pay its huge external debt. This unusual increased production opened the prospects of demand for agricultural labour large farms but it was not enough to compensate for a population growth rate at over 2.5 percent. In combination with this population growth, unequal land access produced a surplus in agricultural unemployment. Daily wages demonstrated a downward slide with rural and urban industrialisation lacking sufficient capability to employ the excess labour. Thus, economically impoverished millions of poor agricultural labourers and landless peasants migrated to urban areas such Mamila already overburdened with shanty towns, and millions others moved to the least productive and often ecologically vulenerable territories such as steep hill sides.[25] In these upland areas the settlers used fire to clear the forested land for agriculture and to get charcoal for the cities. Thus, the slash and burn farming and procuring charcoal for the cities with little wherewithal to protect the fragile ecosystem resulted in horrendous environmental damage particularly water erosion, land slides and changes in the hydrological cycle.[26] This damage caused a dismal fall in food production, and encouraged settlers to clear new forested land, thus ensuring massive land degradation. The often sought for marginally fertile land is no longer in sight in many places and economic conditions remain as desperate as before for many peasants.[27]

Significance of Context

Not in all cases environmental scarcity or degradation has been a direct cause of conflict or violence. In cases of scarcity of renewable resources like water, fish, the conflict may be direct or where the environmental cause is found to be dominant. In case of environmentally induced migration the conflict is in association with social, political or economic contexts. These contexts are to be taken note of while studying the environmental causes of conflict. For example, some find behind the 1994 Rwandan genocide, the rapid population growth, severe land scarcity and the resulting food shortfalls- as the key factors. According to Washington Post in its editorial comment: while the Rwandan civil war was "military, political and personal in its execution", a key underlying cause was "a merciless struggle for land in peasant society whose birth rates have put an unsustainable pressure on it."[28] On close scrutiny , the genocide in Rwanda was nothing but " a struggle among elites for control of the Rwanda state. Land scarcity played at most a peripheral role by reducing the regime legitimacy in the countryside and restricting alternatives for elite enrichment outside of government."[29]

While putting a premium on contexts one should not swing to the opposite extreme saying the causal role of environmental scarcity is entirely subordinate to that of contextual factors.To some skeptics, scarcity is wholly a result of political, economic and social factors, such as failed institutions and policies and needs a little independent attention.[30] Homer Dixon refutes this argument on the following grounds :[31]

First, environmental scarcity is not only influenced by social factors like institutions and policies, it can itself affect these institutions and policies in harmful ways. For example, the Mauritanian government built a series of dams along the Senegal river to avoid food shortages and drought, as a result the land prices hiked in these areas, given the critical land scarcities elsewhere in the regions. Being controlled by Moors of Arab region, the government in order to get profit modified the existing laws relating to land ownership even abrogated the traditional right of black Mauritanians to farm, herd and fish along the river. This harmed the interests of a large sections of the country's population.

Second, the environmental scarcity is not wholly a result of economic, political and social factors. It is also partly a function of a particular physical characteristics of society's surrounding environment. These characteristics are, in some respects independent of human activities. For example, the depth of soils in the Filipino uplands prior to land clearing and features that make Israel's aquifers vulnerable to salt intrusion and Bangladesh's physical environment are physical givens of these environmental resources. Third, one environmental scarcity is irreversible, then the scarcity is an external influence on society. (Haiti's top soil washes away into the sea). Even if new policies are undertaken to remove the underlying political and economic causes, the scarcity remains irreversible and remains a continuous burden on society. For example, chronic instability in Haiti and insurgency and violence elsewhere around globe such as in Filipino cannot be understood and remedied unless and until the independent causal role of environmental scarcity is understood and accepted not subordinate to contextual factors.

On the basis of preferential values, beliefs, social and political institutional behaviour, domestic and international arguments can be advanced in favour of not resorting to nuclear brinkmanship. These contexts are no doubt important and become effective in avoiding a nuclear holocaust but these do not in any way diminish the independent importance of possession of nuclear weapons on the part of a state as a security threat. Similarly, in some cases for example as argued, adaptation is a measure that people undertake by suffering a hardship in the form of starving, or death hazards in response to environmental stress, thus avoiding the slide into the loop of environmental conflict. Here, adaptation may not have resulted in a conflict as conceptualized by the dominant environmental conflict paradigm school of thought, but the threats posed by environmental degradation, say due to global warming or green house effect – the result of a policy and function of some governments remain irreversible, because scientific study on the causal role of nations on the region basis and their impact on nation wise has remained to a greater extent uncertain. For example, a scientific study is yet to be made out.

pointing out what quantity of emission of CO_2 from US to the atmosphere causes what amount of global warming, effects on agriculture, health, food etc on how many poor countries. This kind of scientific uncertainty, and study in general has not been able to muster and organize the affected people politically against the cause nations. Even if there is adaptation or adjustment or surrender the threats to the people from the environmental degradation continues unabated. Thus environmental security is a broad and comprehensive spectrum which includes both the environmental causes of conflict and the adaptive measures, so that a review of policies, values and social and political preference of institutions can take place to ensure human security not in terms of adaptive suffering or resorting to conflict for resolution but in terms of an affirmative, positive proactive policy, renewed vision and new planetary ethics.

Scenarios of Environmental Scarcity

Cropland Scarcity

At present total global cropland stands at about 1.5 billion hectares which is about twice the area of the lower forty eight states in the US. Total arable land on the planet including current and potential cropland amounts to 3.2 to 3.4 billion hectares. Most of the best land has already been exploited and the rest is less fertile, rain starved and not easily irrigable and infected with pests. A country is described as land scarce when 70 per cent or more of arable land is under production. In Asia the most populous continent, the per cent of all arable land stands at 82. Though the percentage is lower in Africa and Latin America, the poor quality of much of remaining land is indicative of the fact that the previously high rates of cropland expansion cannot be continued.[32] Land scarcity has already reached a critical proportion in some areas such as Sahel, semi arid region of East and South East Africa, Southern Mexico, Andean highlands and Central America. World's largest untapped agricultural land lies under the rain forests of Congo and Amazon basins-two of the world's main repositories of biodiversity. Soils here being thin and readily eroded by tropical rains are generally fragile and vulnerable to abrupt degradation.

During the 1980s, cropland grew at just 0.26 pre cent a year in developing countries less than half the rate of the 1970s. More importantly arable land per capita plummeted to 1.9 per cent a year.[33] According to an estimate by experts in the absence of a major increase in the arable land in developing countries, the world average of 0.26 hectare of cropland per capita will come down below 0.20 hectares by the year 2025 in view of the growing rate of world population growth. The benchmark to plumb the severity of land scarcity remains at 0.07 hectares of land per capita, though contentious because they depend on assumption about prevailing resource use technologies, capital availability, and other inputs to production. In 1990 Egypt was well below the threshold, at about 0.055 hectares per capita, and by 2025 population growth alone will push the countries such as Bangladesh, Kenya and Israel far under the threshold. The inequalities of land distribution existing in every society mean that the weaker and marginal groups will be subjected to much deepening scarcity. As is revealed, tens of millions of marginalized peasants in the developing world already subsist on very small, often fragmented plots of low quality land. It is estimated that more than 70 per cent of all rural households in these developing countries are either landless or nearly landless. According to Jeffrey Leonard, in the late 1980's "935 million rural people live in households that have too little land to meet the minimum subsistence requirements for food and fuel."[34] This excludes China which at present could add as many as 100-200 million people to the category.

Tracts of land around the globe every year are being lost to a congeries of problems, including encroachment by cities, erosion, depletion of nutrients, acidification, compacting, and salinization and waterlogging from over irrigation.[35] A known conservative in assessment of environmental damage, Vaclav Smil estimates that 2 to 3 million hectares of cropland are lost annually to erosion, perhaps twice as much land goes to urbanization and at least 1 million hectares are abandoned because of excessive salinity. In addition, about one-fifth of the world's cropland is suffering from some degree of desertification. Taken together, he concludes the planet will have lost about 1 million hectares of arable land between 1985 and 2000 equal the area of California and Texas combined.[36] A study made by the International Soil Reference and Information Center and the United Nations Environment Programme in association with over 250 soil and environmental scientists in 1990, gives a general impression about the state of land degradation in the world according to which, 38 per cent of world's current cropland, 65 per cent in Africa, 51 per cent in Latin America and 38 per cent in Asia are affected by land degradation.[37]

Deforestation

The study made by the UN Food and Agriculture Organisation (FAO) suggests that the already high rate of global tropical deforestation of the early 1980s has gone up again. In the early 1980s, the world was losing 11.3 million hectares of tropical forest annually. By 1991 the FAO put the figure at 17 million hectares.[38] Between 1960 and 1990 Asia is estimated to have lost nearly a third of its tropical forests, while Latin America and Africa lost almost a fifth.[39] At present, Brazil, Indonesia and Malaysia these three countries alone account for over half of the world's tropical rain forests.[40] Other countries experiencing high rates of deforestation include Costa Rica, Cameroon, India, Myanmar, the Philippines, Thailand, and Vietnam.

It is not that in developing countries only deforestation has taken place. The impression has been created to show that it is the developing countries and their growing population that have been responsible for the present heightened rate of deforestation. The fact is that after decimating their forest cover, the industrialized countries now shift the responsibility and burden to the developing countries to protect the forest cover. So use of phrases like "common heritage of mankind", and "global commons" is used to make forests in developing countries as carbon sinks. According to Gregg Marland, indeed out of the 800 million hectares of woodlands, about 300 millions hectares – well over a third – has taken place only in the industrialized countries.[41]

Water Scarcity

Heightened competition for fresh water poses a major threat to human security the world over. It could well prove to be the major source of conflict in the 21st century, not unlike oil in the present century.[42] Water depletion or degradation is supply induced. Water scarcity crisis is gravest when viewed in terms of rising demand. Reduced and degraded water quality has a direct impact on human health, about 250 million new cases of water borne diseases are reported every year causing about 10 million deaths.[43]

In 1999 human beings were estimated to draw about 4,250 cubic kilometers of fresh water a year from various sources returning about 1,700 cubic kilometers often in polluted conditions.[44] This consumption is growing at a rate of 2 to 3 per cent a year. The annual quantity of water available from rivers is probable closer to 40,000 cubic kilometers.[45] Though this suggests water supply is in abundance, great difference in water availability is marked among regions. In a number of arid developing countries burgeoning population growth threatens to reduce water availability below 1,000 cubic meters of water per capita per year, a benchmark regarded by experts as about the minimum necessary for an adequate quality of life in a moderately developed countries.[46] Such average can further slide down, as such are deceiving, as large disparities in water access that exist in all societies are kept hiding in working out this average. In such a situation of water, average dwindling below 1000cubic meters in a developing country a bulk of its population is likely to confront serious water scarcity. Then this would result in restricting irrigated agriculture to a greater extent, and slum dwellers and urban squatter settlements will have insufficient water for basic hygiene and rural families will have to walk a long distance to fetch water from streams and ponds.

The following table[47] identifies region where water scarcity is likely to take place in the next decades as a result of population growth. Green house induced climate change is likely to affect these estimates because it would shift precipitation patterns and therefore the overall water supply.

Water Availability in 1995 and 2025

Country	Per Capita water availability 1995 m^3 / person & year	Projected Per Capita Water Availability 2025 m^3/ person
Africa		
Algeria	527	313
Burundi	594	292
Comorous	1667	760
Egypt	936	607
Ethiopia	1950	807
Kenya	1112	602
Libya	111	47
Malawi	1933	917
Morocco	1131	751

(Contd.)

(Contd.)

Country	Per Capita water availability 1995 m^3 / person & year	Projected Per Capita Water Availability 2025 m^3/ person
Rwanda	1215	485
Somalia	1422	570
South Africa	1206	698
Tunisia	434	288
The Americans		
Canada	98,667	79,731
United States	9,277	7,453
Barbados	192	169
Haiti	1,544	879
Asia/ Middle East		
Bahrain	162	104
Cyprus	1,208	947
Iran	1,719	916
Israel	389	270
Jordan	318	144
Kuwait	95	55
Oman	874	295
Qatar	91	64
Saudi Arabia	249	107
Singapore	180	142
Yemen	346	131

Despite the uncertainty introduced into the estimates of water scarcity by the green house gases emitted into atmosphere by the industrialized nations, the Middle East and parts of Africa are likely to be affected severely by water scarcity. In these countries, population are expanding rapidly and water is already extremely scarce.[48] According to a study the number of people living in countries severely affected by water stress or chronic water scarcity in 1997 stands at 430 million. This number is expected to rise to 3 billion or over a third of the planet's population projections.[49]

Dispute over river water becoming scarce day by day with growing population it is held would become more frequent among or between nations in coming decades.

Decline in Fish stocks

There is widespread evidences – from Maine to Newfoundland, to Southern India- of regional over exploitation of fish. Many fisheries are reported to have suffered from eutrophication, coastal pollution and destruction of mangrove swamps – the breeding grounds of prawns. Tropical fisheries especially in South East Asia are being ravaged by a combination of dynamite and cyanide fishing. These large swings in productivity are related to global climate changes, in particular El Nino oscillation in the South Pacific.[50] As early as 1987, the FAO commented: "The time of spectacular and sustained increases in fisheries catches is over.... Almost all important stocks of demersal species are either fully exploited or overfished. Many of the stocks of more highly valued species are depleted. Reef stocks and those of estuarine / littoral ones are under special threat from illegal fishing and environmental pollution."[51] In a 1995 report, the FAO concludes that about 69 per cent of the World's conventional species are "fully exploited, over exploited, depleted, or in the process of rebuilding as a result of depletion.' Further it notes that, " this situation is globally non-sustainable and major ecological and economic damage is already visible".[52]

Total world sustainable fish both marine and fresh water fish production which has quintupled since 1950, from 20.75 to 101.27 million metric tons in 1993 stands at 100 million metric tons.[53] To maintain current per capita human fish consumption the world will have to produce about 91 million metric tons of food fish by the year 2010.[54] Of the approximately 86 million metric tons of fish caught from marine and inland fisheries, about 57 million metric tons went to human consumption. It is unlikely that the quantity of food fish caught from marine and inland fisheries will exceed 60 million metric tons annually.[55] The remaining 31million metric tons will have to come from aquaculture which means doubling of aquaculture production from its 1993 levels of 16 million tons. This seems highly improbable. In such a situation, increasingly sophisticated fishing fleets may be used to push fish production beyond sustainability to meet the fish requirement of the people. This is likely to affect the marginal communities and small-scale fishers along the coasts of developing countries who entirely depend on traditional fishing.

In addition to increasing human numbers overfishing using high tech electronics to track fish and competition among fishermen to catch as quickly as possible without any regard to the long term maintenance of the fisheries are one of the primary reasons for their decline.[56] A big contributor to the pollution of ocean and seas is the growing urban population, of which one third live within 60 kilometers of the coastline. Heavy metals, sewage, indiscriminate use of fertilizers, solid wastes from industries, run off from agriculture, and mining of sands and coral reefs have seriously affected sea life, overfeeding algae and causing it to bloom in a rapid growth that uses up the oxygen fish need to breathe.[57] The stratospheric ozone hole radiating more ultraviolet rays to the earth affects the sea life causing an increasingly mortality of fish. Global warming is very much likely to affect the sea life because it could alter water currents, which circulates heat and vital nutrients. The photosynthesis and growth of phytoplankton at the lowest link of the food chain, have decreased by as much as 20 per cent near the surface of Antarctic waters- at the surface where most marine growth and reproduction takes place.[58] Not only decline in fish catch but also their contamination has been a serious problem for fish consumers.

Loss of Biodiversity

Another important casuality of human activity inflicting damage on Earth's renewable resources is the priceless bio diversity. The estimates of number of species on the planet cover a wide range, from 5 to more than 30 million. The current best estimate known to science is 1.75 million[59] and only 13 per cent of the species on Earth has yet been described.[60] The tropical forests are the largest repository of genetic information, the majority of it contained in insects and microbes not yet identified. The world's biodiversity is a priceless resource for the development of new crops, medicines and a wide range of industrial products from paints to lubricants. With the destruction of tropical forests, the earth is robbed of most of its genetic information. Industrialization of agriculture, planting of trees and croplands to obtain high yields of products in grain and woods have resulted in simplification of ecosystem which in turn has reduced ecosystem stability and the availability of services such as soil nutrients and pest control.[61]

With the rapid destruction of tropical forests and other species' habitats around the Earth, both animal and plant species are vanishing at an extraordinary speed. According to an entomologist, the global loss from tropical deforestation alone could range from four to six thousand species a year, a rate ten thousand times greater than the natural background rate of extinction prior to the advent of human beings on earth.[62] The Global Biodiversity Assessment similarly concludes that "recent rates of deforestation translate into a rate of extinction of about 0.25 per cent per annum."[63] Though this is a crude estimate "the current extinction rates are dramatically higher than background extinction rates."[64] Urban growth, population growth, pollution, human predation and loss of habitats to loggers all contribute to the crisis. If climate change stress is an indication, then "an estimate of a 25 per cent reduction in planetary biodiversity in the next one hundred years is plausible. Such a loss would rival four of the five previous mass extinctions on earth. From both a moral and practical point of view, it could be the single greatest calamity human beings inflict on the planet."[65]

Global Warming

A world heavily dependent on carbon based fuels for economic growth and development finds that their use causes a build up of carbon dioxide in the atmosphere which heats up the planet like a garden green house. It is the western pattern of development based on fossil fuel based industrialization that has been solely responsible for accumulation of CO_2 in the atmosphere. Industrialized countries owe their current prosperity to years of historical emissions which have heaped up in the atmosphere since the very inception of industrial revolution, and also to a high level of current emissions. The use of these fossil fuels and hence carbon emissions are closely linked to economic growth and life style. Every human being contributes to the carbon dioxide concentration in the atmosphere, though the person's life style decides the amount that is emitted. The more prosperous a country's economy and higher its per capita income, the higher its fossil fuel consumption for power generation and transport, and the higher its GHG emissions. The following tables[66] show a comparative statement of emissions of CO_2 to the atmosphere by both industrialized and the developing countries:

Percentage and distribution of net emissions of Green house gases (GHGs) by industrialized and developing countries.

Country	WRI formula	CSE calculation
United States	17	27.4
Soviet Union and Eastern Europe	16	17.6
Western Europe	14.3	11.9
Japan	3.9	2.5
Other industrialized countries	1.4	7.5
India and China	10.3	0.6
Brazil	10.3	18.2
Other developing countries	26.7	14.3

Comparative figures of total emissions of GHG s of WRI's top ten emitting nations as calculated by CSE (million tonnes of carbon equivalent)

	WRI	CSE	WRI	CSE
US	1000	1532	1368	836
Soviet Union	690	730	882	842
West Germany	160	155	144	149
United Kingdom	150	132	160	178
Japan	220	140	202	281
France	120	69	121	173
Brazil	610	1017	826	418
China	380	32	793	1141
India	230	0.7	643	871
Indonesia	140	9.5	240	370

The WRI report was widely quoted to point out that developed and developing countries have contributed almost equally to the global warming problem. On the contrary, WRI's own data and methodology showed, when reinterpreted by CSE, that developing countries as a group contributed to only 33 per cent of the net emission in 1987. This was when the unused emissions of developing countries were traded with all countries including industrialized countries on an equal per capita basis. But when the unused emissions were traded only with other developing countries which were emitting in excess, the industrialized countries' share increased to 91 per cent while that of the developing countries dropped to nine per cent.[67]

In an analogy to national financial debt, Kirk Smith calls these cumulative emissions the 'natural debt of a nation'. Just as the nations take on a financial debt over the years in order to grow faster, they also take on a natural debt for the same reason. In the latter case economics borrow the assimilative capacity of the environment by releasing waste gases faster than they can be removed naturally. His calculation shows that the developed countries are the holders of the largest natural debts. 260 t/capita for the US compared to six t/capita for India in 1990. The United States seems to be responsible for some thing like one-third of the global natural debt, mainly because of its heavy use of fossil fuels starting early in the century. If annual US per capita emissions were to remain at their present value 5.2 t/y and India's per capita emission 0.22 t/y were to continue, India would reach the level of one t/y in the year 2024- a level that was surpassed by the US well before 1900. During the 35 years between 1990 and 2024, the per capita Indian natural debt would have quadrupled six to 24 tonnes / capita in India while an America's per cent capita natural debt would have increased by 70 per cent.[68] From the standpoint of global warming, however the US would have added ten times as much debt… In terms of total atmospheric accumulation, India would be responsible for less than five per cent of that of the US.

Science has been used in the past by the industrialized countries to show that not only the effects but the causes belong to all. The methane gas emitted from paddy cultivation in South particularly in India and Thailand is highlighted to gloss over the historic responsibility that lies with the industrialized North to stop emitting CO_2 to the atmosphere right now, by making no distinction between the survival emission of the South and the luxury emission of the North.[69] In all international negotiations relating to climate change US insistence on meaningful participation of the South exposes American double standard and negation of equity while propounding this in dealing with the task of cleaning the atmosphere but not in case of utilising the atmosphere which the industrialized world in a sense colonized it to the peril of the South. Since emission of CO_2 to the atmosphere is closely related to economic development and life style, the high level accumulation of CO_2 by industrialized North suggests that it is the inordinate life style and consumption pattern of the North people and the continuing persistence of development paradigms based on fossil fuels in the North, that has been responsible for global warming and its consequent effects on human life. Thus, the climate change negotiations have already been sucked into the deep water politics of their respective national interest defined in terms of economic growth and consumptive life style. America's rejection of Kyoto Protocol which it signed to reduce CO_2 to the agreed level, is on the same ground that reduction of its level of CO_2 would mean acceptance of a lower standard of living on the part of its citizens, and a challenge to its existing life pattern and above all winding up of some fossil fuel based industries which America is not prepared to do being influenced by the powerful coal automobile industrial lobby in US.[70]

There has been a welter of scientific uncertainty about positive and negative feed backs that may accentuate or diminish human perturbations of the global heat balance. And also very little research has been conducted on the possible impacts of climate change on different countries and regions. However over the last few years, a consensus has developed among the experts: assuming no major changes in the trend of human emission of green house gases, We will likely see a global warming between one and three and one-half degrees Celsius by 2100, with a best estimate of two degrees warming.[71] This seems not to be a much of an increase to create panic in view of the fact that the earth has warmed only about five degrees since the coldest period of ice age, around 18,000 years ago. But what is of concern to mankind is the predicted rate of warming which during the next hundred years will " probably be greater than any seen in the last 10000 years.[72] Already global mean surface air temperature has risen by 0.3 and 0.6 degrees Celsius since the late nineteenth century, and the last decade has witnessed a number of the warmest years on record.[73] There is a growing consensus among the scientists about 'a discernible human influence on global climate."[74]

According to the second report of the Intergovernmental Panel on Climate Change (IPCC), global mean surface air temperature(the Earth's surface temperature) has increased between 0.3-0.6° C. since the late 19th century. It estimates that global mean surface temperature would be about 2°C above pre industrial levels by the year 2030, and about 4°C above pre-industrial levels by 2090. Global sea level has risen by 10-25 cm over the past 100 years. It is projected to rise between 9cm and 29cm by 2030 and 28cm and 96cm by 2090.[75]

The consequences of global warming according to this report are as follows:[76]

- Changes in weather patterns – more floods and droughts in some areas, and far reaching effects on agriculture and forestry.
- Scarcity of drinking water and sanitation–millions would be affected.
- Food crisis – decline in food production in the tropics and subtropics, despite steady global production.
- Flooding– rise in sea levels would submerge small island states and low lying deltas, such as in Bangladesh, Egypt and China. Millions would be driven out of their houses as refugees.
- Deaths and diseases – heat stress and vector borne diseases would claim more lives as the tropical habitat of insects expands northwards.
- Loss of biodiversity – species adapted to cool climates would become extinct as habitats disappear and they are unable to migrate.

Ozone Layer Depletion

The ozone issue was first noticed worldwide in1985 when a British scientist reported that a hole in ozone layer had been occurring over Antarctica every spring since 1979. The scientists who then sprang into action confirmed that the ozone layer was being depleted far more rapidly than had been forecast and the cause of this being the presence of human made chemicals chloroflurocarbons (CFCs) in the upper reaches of the stratosphere. The ozone layer provides a fragile shield for the earth against harmful ultraviolet radiation (UV-B) from outer space. Ever since its discovery, this hole has been growing steadily. In the Antarctica spring of 1998, the ozone hole reached the largest ever covering an area of 2.6 billion hectares more than two and a half times the area of Europe, and even covered some populated areas of the Southern hemisphere. In 1999, the ozone hole had an average of 2.2 billion hectares, only slightly smaller, attributed rare to climatic changes than any significant reversal of the phenomena. While no hole has appeared elsewhere, the Arctic spring has seen the ozone layer over the North Pole thin by up to 30 per cent, while the depletion over Europe and other high latitudes varies between 5 per cent and 30 per cent.[77] Scientists have warned that if this hole is not sealed in the sky, it could lead to enormously increased penetration of ultraviolet rays which would increase skin cancers, cataracts and depress the human immune systems. Satellite observations show that since 1972, the amount of UV radiation has increased an average of 6.8 percent per decade in the northern hemisphere and 9.9 per cent per decade in the South.[78] According to a study, on average a 1 per cent decrease in stratospheric ozone produces about 1.6 per cent increase in the incidence of carcinogenic UV radiation on the surface of the earth, which in turn produces about a 2.7 per cent increase in nonmelanoma skin cancer rates.[79] A ten per cent reduction in ozone is predicted to result in 24 per cent increase in incidence of non-melanoma skin cancer, a 30 per cent reduction will lead to a doubling and a 50 per cent reduction in quadrupling of such cases. Dramatic increases in all forms of skin cancer have been reported worldwide over recent decades but rates of non-melanoma skin cancers have increased so significantly that it has been labeled "a quiet 20th century epidemic."[80] What is of greater concern is the deleterious effects of increased UV radiation on crops, forests, amphibians, ocean phytoplankton and human and livestock health.[81]

The problems of rising cropland scarcity, tropical deforestation, growing fresh water scarcity and depletion of fish stocks when interact with the aforementioned effects of climate change, will very much threaten the well being of the many developing states. The global inequities in CFC use are enormous. The global consumption was and remains lopsided with the developed world in 1986 using more than 85 per cent of the total. The United States alone consumes roughly 30 per cent of the World CFC total followed by Japan and Europe. The developing countries' share is miniscule with India and China together consuming about 2 per cent.[82]

The Montreal ozone protocol signed in 1987 by the industrialized nations, instead of having set out a system in which a global common resource is shared and managed equitably, with equal rights for every citizen, the treaty operates within the context of existing inequities.[83] While this agreement suited the economic interests of the North, where the major producers of CFCs were based, it was unfair on the South because it froze existing and unequal consumption patterns. Its 1987 version therefore gave richer countries the right to pollute more. Even after this version was implemented, industrialized countries were allowed to have a higher per capita consumption of ozone depleting substances than developing countries. While all developing countries had to restrict their use to 0.3 kilogram per capita, the per capita consumption for US for instance was at 0.85 kg per capita.[84] Its compliance mechanism is only effective against developing countries, which depend on the protocol for funds. It does not cover any industrialized country's commitments, and any efforts to hold industrialized countries responsible for their actions have failed.[85] This is a flaw that needs to be addressed in view of the damaging effects of ozone hole on human life.

In spite of a decline in production and consumption of CFCs because of the industrialized countries abiding commitments to the treaty, the challenges remain. In the words of Topfer, contrary to popular impression, the protection of ozone layer is far from assured at present."[86] The Montreal Protocol envisages peak ozone depletion around 2000-2005 following which a steady improvement culminates in full recovery around 2050. But a 1998 study by NASA scientists suggests that this production might have been too optimistic. According to NASA, the depletion of Arctic ozone will not peak until 2010-2019, between 10-20 years later than expected. It also warns of likely development of a severe Arctic ozone hole, tripling the levels of UV radiation hitting Northern countries perhaps due to the impact of global climate change on the ozone layer.[87]

Energy Consumption

During the past two decades, humankind's primary energy consumption has gone up steadily by 2 to 4 per cent a year. Today it stands at 8.8 billion metric tons of oil.[88] In many developed countries, per capita energy consumption is thirty or more times than in developing countries. Currently oil makes up 38 per cent of global commercial energy consumption; coal 30 per cent, natural gas 20 per cent, hydropower about 7 per cent; and nuclear power 5 per cent. As accessible and cheap petroleum reserves are depleted, these percentage will change. Around 60 per cent of the world's population depend on non commercial sources of energy for cooking and heating such as fuel wood, twinges, branches and straw. These people mostly living in Africa, Indian subcontinents and China will face scarcity of these resources as the forests are getting degraded day by day, they have had to go miles daily for fetching these fuels.[89] As pointed out by Dixon, "the level of global energy consumption in coming decades will depend on a wide array of factors, including whether energy prices are adjusted to reflect the costs of pollution and global warming to our societies, whether new technologies allow more efficient production and use of energy, and whether international agreements are achieved to restrict releases of carbon dioxide."[90]

In 1993, the World Energy Council (WEC) published four scenarios for primary energy demand, each using different assumptions about economic growth, energy intensity of production, energy efficiency, technology transfer and institutional improvements. The predicted energy demand in 2020 for these scenarios ranges from 11.3 to 17.2 billion metric tons of oil equivalent.[91] A 1992 study by the MITRE Corporation suggests that global commercial energy demand in 2025 will be about 13 billion metric tons of oil equivalent, which is roughly in the middle of the WEC range. This study also estimates that oil shortage will begin around 2030 or sooner, because of the depletion of oil resources.[92] In 1995, the WEC and IIASA published a joint report that extended the time horizon of 1993 WEC estimates to 2050 .They judged that world economic output would increase by 3-5 times by 2050 and that global demand for energy services could rise by as much as ten times. But because energy intensities will continue to decrease, primary energy requirements would rise by only one and a half to three times (to 14 to 25 billion metric tons of oil equivalent)[93].

What gets clear is the realization that energy security will not come with more drilling. According to one study, "One of the ironies at the turn of the century is that, in an age when the pace of technological change is almost overwhelming, the world will remain dependent, out to the 2020 at least, essentially on the same sources of energy-oil, natural gas, coal-that prevailed in the twentieth century."[94]

Recourse to nuclear energy appears not the viable alternative. Expansion of nuclear power is definitely not worth the high level waste it would generate. The Kyoto Protocol cannot be used to promote nuclear energy. Nuclear energy is not clean till a way is found to deal with nuclear waste.[95] Any strategy that seeks to obtain least cost carbon emission reduction options will inevitably focus on improving energy efficiency in the carbon energy sector. But will energy efficiency within the fossil fuel energy scenario help solve the global warming problem? Dennis Anderson, a senior adviser in the World Bank's industry and energy department, warns :

"It will not be possible to prevent the accumulation of carbon in the atmosphere unless non-carbon (or non-net carbon emitting) alternatives become available. Improving energy efficiency will help and is important for economic as well as environmental reasons, but it will not prevent carbon accumulations from growing exponentially or indefinitely, so long as carbon emissions from the burning of fossils fuels exceed 2-3 billion tons per year- the current estimate of the 'natural' net rate of absorption of carbon by the earth's oceans and land masses. Presently, the rate of emissions is around 6 billion tons per year, and emissions are growing almost in direct proportion to world energy demand; it is conceivable that emissions will exceed 10 billion tons in 20 years, and 20 billion tons in 50 years - and this would be in an energy efficient world."[96]

If the world waits for a large part of its oil and coal resources to be exhausted before the shift to zero carbon energy transition occurs, which will not be before the 22nd century, then the certainty of serious climate change occurring becomes inordinately high almost definite. According to a major study by WEC and IIASA (International Institute for Applied Systems Analysis) the gross global carbon emissions from fossil fuel combustion could rise from 6.23 gtC in 1990 to anywhere between 5.93 gtC to 16.00 gtC in 2050 and per capita carbon emission from 1.18 tC in 1990 to anywhere between 0.59 tC to 1.59tC. Gross carbon emissions of OECD countries could rise from 3.03 gtC in 1990 to anywhere between 4.2 gtC to 9.02 gtC in 2050.[97] Despite focus on renewable sources of energy, the renewable are still unable to penetrate the market. For example, despite 30 years of public discussion and environmental support, renewable energy technologies have failed to emerge as a prominent component of the US energy infrastructure. The Natural Resources Defence Council, a prominent public interest group in US, said, the Bush-Cheney energy plan is the culmination of a process that hinged on cosy business connections, secret deals and oil and energy industry campaign contributions.[98] The biggest obstacles in the way of renewable technologies today are low fossil fuel prices and fossil fuel subsidies in many countries; declining public sector research and development, and plummeting private sector research and development as the deregulation of energy markets increasingly focuses attention on short term returns.[99] In the absence of focusing on renewables as the primary objective of national energy policy by the developed world since they owe a historic responsibility of having wrought a drastic climate change in terms of global warming, sea level rise and other green house effects by emitting maximum amount of CO_2, the continuous focus on carbon based energy will further intensity the climate change with diabolic consequences on the developing world.

Population Growth

The above environmental scarcity of resources is not always the result of a natural process but often the result of an interaction with the population. Population size can be a key force driving supply induced scarcity. In combination with prevailing social structure, technologies, and consumption patterns population growth makes environmental degradation and depletion worse.[100] Conflicting views on population and the environment exist. The Malthusian concept of population as a central factor can be considered. The left of the centre radicals stress on poverty, inequality and over consumption by the North, or technology as the causes of environmental degradation. The right wing radicals ensconce their faith in the unlimited capacity of human ingenuity to respond to the need of the free markets which would enable it to resolve the problems of environmental decay. Whereas population explosion has not been a central concern in the North, it has become predominantly a central concern in the South.

More conveniently the North has been putting the blame for what they have wrought to the environmental on the South's population explosion in a bid to continue the perpetuation of their economic exploitation of the poor South. On the other hand, it is held that the environmental problems lay more with the high consumption levels in the developed countries than with the supposed population levels in the

South. The ethnocentrism of Western paradigms of development based on fossil fuel based technology, and Western economic structure dominating the entire globe with the South occupying the periphery, have resulted in decimating the environment. The role of North in causing global warming and ozone layer depletion is indubious. South's imitating Western pattern of development and life style as a model of development, exploitations of the renewable resources for repayment of debts or to earn more foreign currencies redounded to the economic prosperity of the North at perils of nature.

The latest United Nations medium projections suggest the total human population will grow from the current 6.0 billion to about 8 billion in 2025 and 9.4 billion in 2050.[101] Although the rate of global population growth has dropped from 2.1 per cent to about 1.3 per cent over the last decades, the human population is still expanding by about 80 million a year. It is estimated that the growth will average about 80 million people a year during the entire four decades from 1980-2020. About 50 per cent of the anticipated growth in the developing world's population will result from the age structure of today's population, especially the large number of girls still to reach the reproductive age which guarantees a much larger population in developing world in future even if all these countries make the best use of the available family planning methods.[102] In such a situation where over 90 per cent of today's population growth is taking place in the developing world, the ratio of poor people to rich is rapidly growing. In 1950 the ratio was about two to one, today the ratio is about four to one and in 2025, it will be almost six to one.[103]

It is also held that the population issue is no longer a concern.[104] The inequal economic structure of the world presided over by the North and the inordinate life pattern and consumption of the industrialized World are often glossed over to magnify the population explosion in the South as the key cause of environmental degradation. As some views, the environmental problems lay more with high consumption levels in the developed countries than with the supposed population growth in the South.[105]

Due to population growth, the surplus population in the South are migrating to cities to form slums and shanties raising the prospects of violence. Questions may be raised whether the highly populous states in

South such as India, China, Pakistan, and Egypt have the required natural resources and economic capability to sustain their populations. As a result, these countries' population will perpetrate a huge toll of their natural environment.

References

1. See for example,Lothar Brock, "Peace through parks: The Environment on the Peace Research Agenda," Journal of Peace Research Vol.28,(40),1991, pp. 408-409

2. Joseph Huber, Eine Sozialvissen schaftliche Interpretation der Humanokologie, Glaeser 1989, p. 65

3. Stephan Libiszewski, "What is an Environmental Conflict", ENCOP, Occasional paper no.1, July 1992, p. 4

4. Gunther Bachler, "Conflict and Cooperation in the Light of Global Human-Ecological Transformation", ENCOP, occasional paper no. 9, October 1993, p. 5

5. Stephan Libiszewski, What is an Environmental Conflict, *op.cit.*, p. 4

6. See for several definitions, David Pearce and Kerry Turner, Economics of Natural Resources and the Environment (Baltimore : Johns Hopkins University Press,1990), pp. 52-53

7. Thomas F. Homer Dixon, Environment, Scarcity , and Violence (Princeton/ NewJersey : Princeton University Press,1999), p. 47.

8. United Nations Population Fund : The State of World Population (Oxford : Oxford University Press, 1992), p. 23.

9. Stephan Libiszewski, *op.cit.*, p. 5

10. Judith Rees, "Resources and the Environment, Scarcity and Sustainability", in Robert Benneth and Robert Estall, ed., Global Change and Challenge : Geography for the 1990's (London/New York,1991).

11. Thomas F. Homer Dixon, Environment, Scarcity, and Violence, p. 15.

12. Homer Dixon, and Valerie Percival, Environmental Scarcity and Violent Conflict : Briefing Book (University College, University of Toronto and America Association for the Advancement of Science,1996), p. 6.

13. Homer Dixon, Environment, Scarcity and Violence, *op.cit.*, p. 74

14. Miriam Lowi, West Bank Water Resources, *op.cit.* ; Miriam Lowi, " Bridging the Divide : Transboundary Resource Disputes and the Case of West Bank Water", International Security, Vol.18, no. 1, 1993, pp. 113-38; Stephen Lonergan and David Books, Watershed : The Role of Fresh Water in the Israeli – Palestinian Conflict (Ottawa: International Development Research Centre, 1994), Miriam Lowi, Water and Power : The Politics of a Scarce Resource in the Jordan River Basin (Cambridge : Cambridge University Press,1995).

15. See Fred Pearce , " Wells of Conflict on the West Bank," New Scientist 130, no.1777 (1 June 1991), pp. 37-38.

16. Miriam Lowi, West Bank Water Resource and the Resolution of Conflict in the Middle East, Occasional paper no.I, Project on Environmental Change and Acute Conflict (Cambridge, Mass : American Academy of Arts and Science and University of Toronto, 1992), p. 34.

17. Natasha Beschorner, Water and Instability in the Middle East, Adelphi Paper 273, International Institute for Strategic Studies(London : Brassey's 1992), p. 14, p. 78.

18. Thomas F. Homer Dixon, Environment, Scarcity and Violence, p. 76.

19. N.Gaan, "Water and Conflict in West Bank and Golan heights: Are Environmental Issues and National Security Linked", India Quarterly, July-December, 2002.

20. Homer Dixon, Environment, Scarcity, and Violence, *op.cit.*, p. 16

21. *Ibid.*, p. 74.

22. See Narottam Gaan, The Environmental Degradation and Conflict : The Case of Bangladesh – India (New Delhi : South Asian Publisher, 1998), pp. 100-107.

23. Jeffrey Leonard, "Overview",in Environment and the Poor : Development Strategies for a Common Agenda(New Brunswick, N.J : Transaction, 1989), p. 5.

24. See Richard Kessler, Rebellion and Repression in the Philippines (New Haven, Conn. : Yale University Press, 1989), p. 18

25. See for full account, Maria Concepcion Cruz et al., Population Growth, Poverty, and Environmental Stress : Frontier Migration in the Philippines and Costa Rica (Washington, D.C : World Resources Institute, 1992).

26. World Bank, Philippines : Environment and Natural Resource Management Study (Washington, D.C. : World Bank 1989).

27. Robin Broad with John Cavanagh, Plundering Paradise : The Struggle for the Environment in the Philippines (Berkeley : University of California Press, 1993).

28. Washington Post, " From Rwanda to Cairo", 25 August 1994, p. A18.

29. Thomas F.Homer Dixon and Valerie Percival, " Environmental Scarcity and Violent Conflict : The Case of Rwanda",Journal of Environment and Development no.3, 1996, pp. 270-91

30. See Ronnie Lipschutz, "Environmental Conflict and Environmental Determinism: The Relative Importance of Social and Natural Factors". in Nils Petter Gleditsch, ed., Conflict and the Environment (Dordrecht: The Netherlands: Kluer, 1997), pp. 35-50. See also Wenche Hauge and Tanja Ellingsen, "Beyond Environmental Scarcity: The Causal Pathways to Conflicts", Journal of Peace Research Vol.35, no.3,1998, pp. 299-317.

31. Homer Dixon, Environment, Scarcity, and Violence, op.cit, pp. 17-18.

32. World Resources Institute(WRI), World Resources 1990-91(New York : Oxford University Press, 1990), p. 5.

33. Nafis Sadik, The State of the World Population 1990(New York : United Nations Population Fund,1990), p. 8.

34. See Jeffrey Leonard, "Overview", in Environment and the Poor : Development Strategies for a Common Agenda (New Brunswick, N.J. : Transaction, 1989), p. 13

35. See Sara Scherr and Satya Jadav, Land Degradation in the Developing World : Implication of Food, Agriculture, and the Environment for 2020,IFPRI Food, Agriculture, and Environment Discussion paper 14(Washington, D.C. : International Food Policy Research Institute , 1996)

36. Vaclav Smil, Energy, Food, Environment : Realities, Myths, Options(Oxford : Oxford University Press, 1987), pp. 223, 230

37. Global Assessment of Soil Degradation, World Map on Status of Human Induced Soil Degradation (Nairobi, Kenya/Wageningen, The Neatherlands : United Nations Environment Programme/International Soil Reference Centre, 1990).

38. World Resources Institute (WRI), World Resources 1992-93 (New York : Oxford University Press, 1992), p. 118.

39. World Resources Institute (WRI), World Resources 1996-97 (New York : Oxford University Press, 1996), p. 205

40. World Resources Institute(WRI), World Resources 1994-95 (New York : Oxford University Press, 1994), pp. 130-35

41. Gregg Marland, "The role of forest in addressing the carbon dioxide greenhouse in James C. White, ed., Global Climate Changes Linkages (New York : Elsevier, 1989)

42. Mikiyasu Nakayama, "Water : The 21st Century's Oil", Work in Progress Vol. 15, No. 2, The United Nations University, Winter 1998,.p. 18.

43. Nicholas Kristof, "For Third World . Water is still a Deadly Drink", New York Times, 9 January 1997, p. 41; See also Linda Nash, "Water Quality and Health", in Peter Gleick, ed., Water in Crisis : A Guide to the World's Fresh Water Resources (New York : Oxford University Press, 1993)

44. WRI, World Resources 1990-91, *op.cit.*, pp.170-171.

45. Thomas. F. Homer Dixon, Environment, Scarcity and Violence, *op.cit*, p. 67.

46. Peter Gleick, "Water in 21st Century", in Gleick ed., Water in Crisis, *op.cit*, p.105. Malin Falkenmark. Fresh," Water : Times for a Modified Approach", Ambio, Vol. 15, no. 4, 1986, pp. 192-200.

47. Tom Gardner Outlaw and Robert Engelman, Sustaining Water, Easing Scarcity : A Second Update (Washington, D.C. : Population Action International , 1997); Homer Dixon, Environment, Scarcity, and Violence, *op.cit.*, p. 68.

48. Malin Falkenmark. " The Massive Water Scarcity Now Threatening Africa –Why Is not it Being Addressed ." Ambio. Vol.18, No. 2, 1989, pp. 112-18.

49. Tom Gardner- Outlaw and Robert Engleman, Sustaining Water, Easing Scarcity, Op.cit; and Malin Falkenmark and Asit Biswas, "Synopsis further Momentum to Water Issues : Comprehensive Water Problem Assessment in Being", Ambio, Vol. 24, No. 6, 1995, pp. 380-82.

50. Food and Agriculture Organisation (FAO), The State of World Fisheries and Aquaculture (Rome : FAO, 1995), p. 2

51. Food and Agriculture Organisation(FAO), World Fisheries Situation and Outlook (Rome : FAO, 1989), p. 6

52. FAO, The State of World Fisheries and Aquaculture , *op.cit.*, p. 6.

53. *Ibid*, p. 47.

54. *Ibid*, p. 5

55. Homer Dixon, Environment, Scarcity and Violence, *op.cit*, p. 70.

56. Michael Satchell, "The Rape of the Oceans ", US News and World Report, 22 June 1992.

57. Hal Kane, "Fish catch is no longer growing", in Lester R.Brown, et.al eds., Vital Signs : The Trends That are Shaping our Future 1993-94 (London: Earthscan,1993), p. 32.

58. Testimony of Susan Weiler, Executive Director, American Society of Limnology and Oceanography, Department of Biology, Whitman College, Walla, Washington before Hearing of Global Change Research : Ozone Depletion and its Impacts, US Senate Committee on Commerce, Science, and Transportation,15 November 1991.

59. D.Wawksworth and M. Kalin-Arroyo "Magnitude and Distribution of Biodiversity", Section 3 in V.H. Heywood and R.T Watson ed., Global Biodiversity Assessment (Cambridge : Cambridge University Press, 1995), p. 111.

60. *Ibid*.

61. S.Naeem et.al., "Declining Biodiversity can Alter the Performance of Ecosystems," Nature Vol. 368, no. 6473, 21 April 1994, pp. 734-37 ; D.Tilman,et.al., " Productivity and Sustainability Influenced by Biodiversity in Grassland Ecosystem", Nature, Vol. 379, No. 6567, 22 February 1996, pp. 718-20.

62. Edward O.Wilson, " Threats to Biodiversity", Scientific American. Vol. 261, no.3, September 1987, p. 111

63. R.Barbault and S.Sastrapradja, "Geneneration, Maintenance and Loss of Biodiversity", in V.H. Heywood and R.T. Watson ed., Global Biodiversity Assessment, Section 4. *op.cit*., p. 198.

64. *Ibid*.

65. Homer Dixon, Environment , Scarcity, and Violence, *op.cit*., p. 72.

66. Anil Agarwal and Sunita Narain, Global Warming in an Unequal World : A Case of Environmental Colonialism, (New Delhi : Centre for Science and Environment , 1991), p. 22-23.

67. Anil Agarwal and Sunita Narain, Towards a Green World (New Delhi : Centre for Science and Environment, 1992), p. 33

68. Kirk Smith, "The Greenhouse index." Environment (Washington D.C. : Heldref Publication, 1991).

69. Anil Agarwal and Sunita Narain and Anju Sharma, Globgal Environmental Negotations I, (New Delhi : Centre for Science and Environment, 1999), p. 20.

70. Narottam Gaan, "Politics of Governance of Global Climate Change : Not on Equity but on North's Interests", India Quaterly,vol. LVIII, No.4, October-December 2001, pp. 89-112 ; See also David L.Levy and Daniel Egan, " Capital Contests : National and Transnational Channels of Corporate Influence on the Climate Change Negotaition", Politics and Society, Vol.26, no.3, September 1998, pp. 337-361

71. J.T. Houghton et.al. Climate Change 1995-The Science of Climate Change(Cambridge : Cambridge University Press,1996).

72. *Ibid.*, p. 6.

73. Willam Stevens, "Global Climate Stayed Warm in 1996, with Wet, Cold Regional Surprises," New York Times, 14 January 1997, p. B9.

74. Houghton et. al., eds., Climate Change 1995, Science of Climate Change, *op.cit.*, p. 4.

75. See Anil Agarwal et.al., Global Environmental Negotiations I, *op.cit*, p. 19.

76. R.T. Watson, M.C.Zinyowera and R.H. Moss eds., The Regional Impacts of Climate Change : An Assessment of Vulnerability, IPCC Working Group II Special Report,1998 ; see also Down to Earth, Vol. 6, no. 15, 31 December 1997, P.30.

77. United Nations Environment Programme 2001, Basic Facts and Data on the Science and Politics of Ozone Protection,Press background, Nairobi, Http:// www.unep.org/ozone/pressBack/ as viewed on August 2, 2001.

78. "Ultraviolet Radiation on the Rise", Scientific American Vol. 275, no. 4, October 1996, p. 28.

79. Gert Kelfkens, Frank de Gruijal, and Jan Van der Leun. "Ozone Depletion and Increase in Annual Carcinogenic Ultraviolet Dose", Photochemistry and Photobiology, Vol. 52, no. 4, 1990, pp. 819-23.

80. Gavin Hayman et.al., Chilling Facts about a Burning Issue : CFC Smuggling in the European Union, London Environmental Investigation Agency, 1997, p. 2.

81. "Environmental Effects of Ozone Depletion : 1994 Assessment", Special issue of Ambio, Vol. 24, no. 3, 1995.

82. Cynthia Pollock Shea, Protecting Life on earth : Steps to save the Ozone layer, Washington D.C. : World Watch Paper 87, 1988.

83. Anil Agarwal and Sunita Narain, Towards a Green World, *op.cit.*, p. 8.

84. Anil Agarwal and Sunita Narain, Anju Sharma and Achila Imchen eds., Global Environmental Negotiation 2 ; Poles Apart, (New Delhi : Centre for Science and Environment, 2001), p. 13.

85. *Ibid.*, p. 80.

86. Quoted in Julian Newman and Steven Trent, A Crime Against Nature, (London : Environmental Investigation Agency, 1998), p. 1.

87. Ibid., p. 2.

88. World Energy Council, Energy For Tomorrow's World : The Realities, the Real options, and the Agenda for Achievement (New York : St. Martin's Press,1993), P.III ; and World Energy Council and International Institute for Applied Systems Analysis, Global Energy Perspectives to 2050 and Beyond(London : World Energy Council,1995), p. 45.

89. Dan M.Kammen, "Cookstoves for the Developing World", Scientific American Vol. 273, no.1, July 1995, pp. 72-75.

90. Thomas F.Homer Dixon, Environment, Scarcity and Violence, *op.cit.*, p. 59.

91. World Energy Council, Energy For Tomorrow's World, *op.cit.*

92. S.W. Gouse, et.al., (MI TRE Corporation) "Potential World Development Through 2100 : The Impacts of Energy Demand, Resources and the Environment", in World Energy Council Journal. (London : World Energy Council, 1992), pp. 18-32.

93. Quoted in Homer Dixon, Environment, Scarcity and Violence, *op.cit.*, p. 60.

94. The Geopolitics of Energy into the 21st Century, Centre for Strategic and International Studies, February 15, 2001.

95. Down to Earth,Vol.11, no. 11, October 31, 2002, p. 36.

96. Dennis Anderson, "Energy and Environment : Technical and Economic Possibilities", Finance and Development, IMF, Washington D.C., June 1996, pp. 10-13.

97. Nebojsa Nakicenovic et al.ed., Global Energy Perspectives (Cambridge : Cambridge University Press, 1998).

98. Down to Earth, 31 October 2002, p. 30.

99. Sivan Kartha, et.al., Meaningful Participation for the North and South, Presented at a Workshop conducted by SEI and CSE on Towards Equity and Sustainability in the Kyoto Protocol, Buenos Aires, 8 November 1998.

100. Richard Bilsborrow and H.W.O Okoth Ogendo, "Population Driven Changes in Land Use in Developing Countries," Ambio, Vol.21,no.1, 1992, pp. 37-45; and Tom Rudel and Jill Roper, "Regional Patterns and Historical Trends in Tropical Deforestation, 1976-1990 : A Qualitative Comparative Analysis", Ambio Vol.25, No.3, 1996, pp. 160-66.

101. United Nations Population Division, World Population Prospects : The 1996 Revision, Demographic Indicators 1950, 2050, Diskettes 1-4 (New York : United Nations,1996)

102. John Bongaarts, "Population Policy Options in the Developing World", Science Vol. 263, no. 5148, 11 February 1994, pp. 771-76.

103. See Nathan Keylitz's review of Alfred Sauvy's article in Population and Development Review Vol. 15, no. 2, 1989, pp. 359-62.

104. Ben Wattenberg, "The Population Explosion is Over", New York Times Magazine, 23 November 1997, pp. 60-63; Nicholas Eberstadt, "The Population Implosion", Wall Street Journal, 16 October 1997, p. A22.

105. Kerstin Kiessling and Nathan Keylitz, "The Population Debate : Urgency of the Problem", Paper presented at the Population Summit of the World's Scientific Academics held in New Delhi from October 24-27,1993: See also Down to Earth Vol. 2, 30 November 1993, p. 15.

5

Social Effects, Adaptative Failure and Ingenuity Gap

Environmental scarcity of renewable resources in its interaction with other factors can cause various social effects on the developing countries. According to various research studies there are five main identifiable social effects :[1]

- Decrease in agricultural production
- Economic decline
- Population displacement or migration
- Greater segmentation of society, usually along existing ethnic cleavages and divides
- Disruption of institutions, especially the state

These social effects often found interlinked have feedback relationships.

Decline in Agricultural Productivity

Food production is crucial to the livelihood of most of the people in developing countries. Any shortfall in food production is likely to affect theirs lives, as they spend half of their income on food. There is a causal connection between environmental degradation or scarcity of resources and agricultural output, as suggested by researchers. Effects of land degradation, water scarcity or degradation, deforestation and global warming on food production are obvious in many cases. Coastal cropland in countries such as Bangladesh and Egypt is vulnerable to storms and cyclones, because as predicted global warming will cause sea level rise and might intensify storms. The green house effects will also change precipitation patterns and soil moisture. Lower precipitation will most severely affect regions that already have degraded and eroded soils, because lost rooting depths makes plants vulnerable to droughts.[2] Laboratory experiments under ideal conditions hold out a better prospects for increased crop yields, efficient use of water and larger and faster growing of plants in a warmer climate. But these studies have ignored the influence on yields of more frequent extreme climate change caused by global warming, increased pest infestations and the decreased nutritional quality of crops grown in a carbon enriched atmosphere.[3]

The impact of deforestation on food production is well known. Logging and clearing of lands accelerate soil erosion, change hydrological cycle, precipitation patterns, and decrease land's ability to retain water during rainy season. The flash floods resulting from this damage irrigation works and fill the irrigation channels with silt and plug the reservoirs. Degradation and scarcities of cropland and water affect agricultural production to a significant extent.

It is statistically shown that during eighties world has witnessed an increase in food production due to introduction of green revolution, new technologies, intensive use of irrigation, chemical fertilizers and pesticides. But what is note worthy is that these statistics have hidden the wide disparities in food availability among and within developing countries, and ignored the socio-economic consequences of green revolution. At the end of 1980's, for example, twenty five poor countries could not provide sufficient food calories for the health, growth and productive work of their population.[4] Because of a combination of production and distributional problems, and disparities in access to natural resources, some 840 million people around the world- about 15 percent of the world's population- currently suffer from chronic hunger.[5] Although both the percentage and absolute number of hungry people have slided globally in the last twenty years, Latin America, South Asia and especially Sub-Saharan Africa have not shown improvement in these statistics. Around the world, 34 percent of the world's children are malnourished, about eleven thousand die everyday from malnutrition or related problems; and tens of millions of children suffer from sickness or learning disabilities because of deficiencies of micro nutrients like iron, vitamin A and iodine.[6]

Technological ingenuities such as the green revolution to meet the problems of growing population and food scarcities have not proved to be panaceas. For example, green revolution introduced into two states of India, Haryana and Punjab by the scientific community in the 1960s raised in the beginning hope of rise in food production, through whether it was irrigation, energy, chemical fertilizers or pesticides. For decades down the line more has brought in more problems. Lost of production has spiralled, but the rate of increase in production has not shown a similar trend. Heavy withdrawal of nutrients from soils through intensive cropping patterns has brought down the levels of micronutrients to a point of deficiency. The cost of production has disregarded the ecological and social cost.[7] Thus, the green revolution has turned a nightmare. The regional scarcities of land and water will be having serious effects on food production in developing countries as these countries' food demands grow up and as they reach the limits of potential gains from green revolution. As projected, increased per capita consumption of grain combined with population growth will increase annual consumption of grain of developing countries from 875 million metric tons in the 1980s to 2.35 billion metric tons by 2030, world consumption will rise from 1.68 billion tons to 3.3 billion tons.[8]

To meet the food demands of the growing population in the face of environmental scarcity of land and water what is needed is to increase food production, which is to come from higher yields. By 2030, grain yields in developing countries will need to average nearly 7 metric tons per hectare. Yet the current world average is only 2.8 tones.[9] and many farmers in countries like India and Pakistan do not reach half this average.

According to a study, "many regions will experience virtually no improvement in food security", although the aggregate global supply and demand picture is relatively good. This study further shows that, "there will be very little reduction in the number of malnourished children" in the developing world as a whole; the number will actually increase in sub-Saharan, Africa and parts of South Asia.[10]

As figured out by Homer Dixon, the possible effects of environmental change on agricultural productivity can be studied.[11]

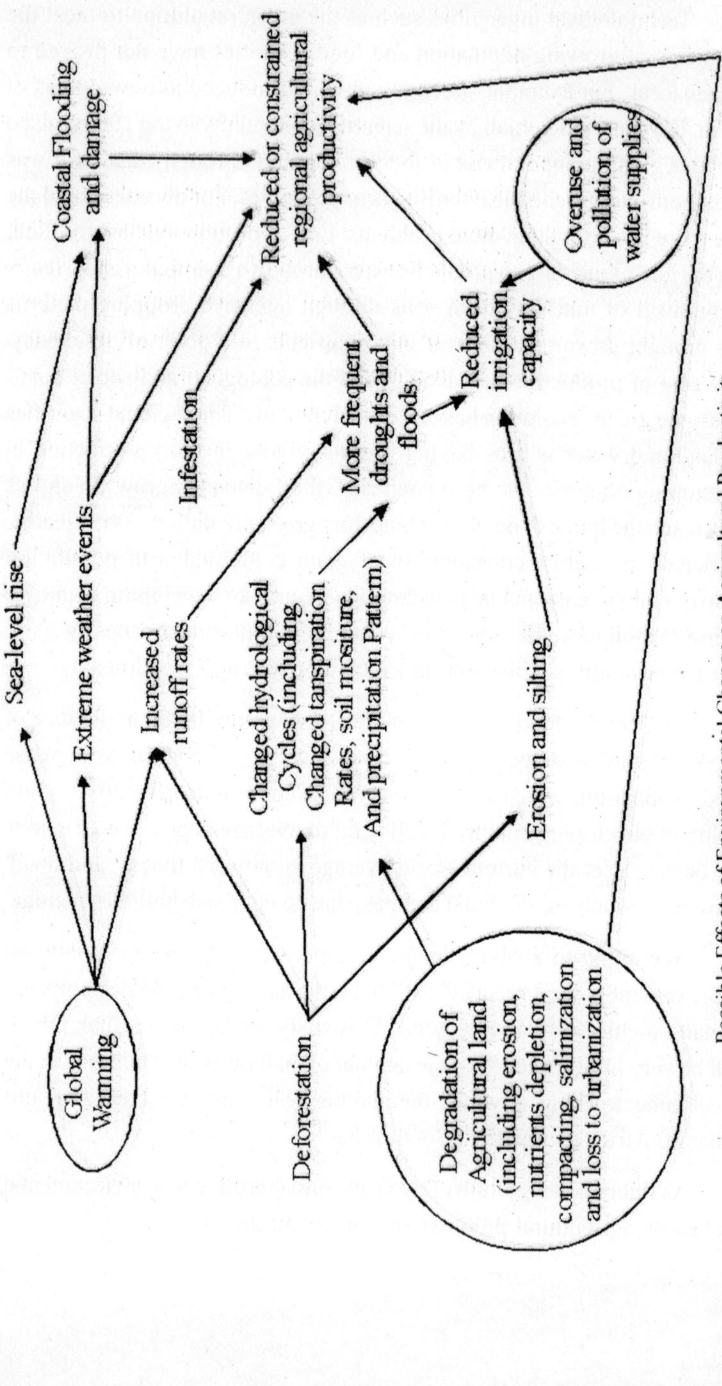

Possible Effects of Environmental Change on Agricultural Productivity

Economic Decline

According to Volker Boge, economic decline may take the following forms due to overuse, and pollution or destruction of renewable resources.[12]

- Decrease in agricultural production
- Insufficient supplies of basic goods (e.g. fuel wood) for the population
- Problems regarding industrial production, transportation systems, traffic etc
- Decline of production for the world market and associated effects
- Production of industrial toxic waste, and waste in general
- Industrial accidents/ catastrophes and their economic repercussions
- Large negative effects on human health

As seen above, environmental degradation or depletion might influence economic productivity directly or indirectly via other social effects, such as changes in agricultural productivity. A great variety of environmental stresses could affect wealth production. Hypothetically, for example, increased ultraviolet radiation caused by ozone hole could eventually raise the incidence of diseases in humans, with serious economic repercussions over the long run.[13] There is abundant evidence of a strong link between other kinds of environmental scarcity and economic productivity. For example, logging for export markets creates short-term economic gain for the exporting country's elite, but the resulting deforestation can affect the economy's longer-term productivity. Increased run off damages roads, bridges and other vulnerable infrastructure, while the extra siltation reduces the transport and hydro-electric capacity of rivers. With forests decreasing day by day, wood becomes scarcer and more expensive, and it absorbs an increasing share of the household budget for the poor families that depend on wood for fuel. The poor families have had to walk a long distance to fetch fuel wood. Since agriculture is the only source of living and wealth generation in developing countries, environmental scarcity and degradation has a deleterious effect on their overall economy. As Per Pinstrup Anderson points out, " in most low income developing countries, growth in the agricultural sector whether through food or non food agricultural commodities, is the most effective and often the only viable lead approach to sustainable economic growth and poverty alleviation."[14] Some experts predict that the economic relief brought by the green revolution within many poor countries will be short lived. Jeffrey Leonard notes: "Millions of previously poor families that have experienced less than one generation of increasing wealth due to rising agricultural productivity could see that trend reversed if environmental degradation is not checked."[15]

Robert Repetto of the World resources Institute noted in regard to the impact of environmental degradation of land on economy: "With erosion, farm output and income have fallen in some regions without major changes in farm practices; other farmers have been induced to change cropping patterns and input use; and in extreme cases, erosion has led to the complete withdrawal of land from cultivation."[16] According to his estimate, erosion in upland Indonesia annually cost the country's agricultural economy nearly half a billion dollars in discounted future income.[17] Gaza offers a particularly stark example of the impact of supply induced scarcities of fresh water on agricultural productivity, on health and, ultimately, on economic development.[18] To buttress his point, Homer Dixon has quoted the Chinese researcheres study of the combined effects of supply induced environmental scarcity on China's economic output.[19]

Migration

One of the consequences of the interaction of the sources of environmental scarcity of resources, is the migration of the people from one region to another region within a state or from one state to another as environmental refugees with destabilizing effects on both domestic order and international security.[20]

In 1985 the United Nations Environment Programme introduced the concept of environmental refugees. They were defined as those people who have been forced to leave their traditional habitat temporarily or permanently because of marked environmental disruption. This is due to reason both natural and anthropogenic that jeopardized their existence and/or seriously affected the quality of their life.[21] So it is ascertained that environmental degradation has been a major and predominant cause of flight or migration of people both intra or interstate. Among a complex web of factors responsible for migration, environmental degradation may act as a precipitating event that "disrupts the normal functioning of the system and thus destroys the capacity of a population to survive under the prevailing conditions,"[22] Environmental degradation in the natural processes by human activities in the face of what Giddens calls the " high consequences of modernity",[23] may collapse the very social systems which provide people with an ongoing sense of ontological security. Not only environmental scarcity of resources but their consequences in the form of global warming, greenhouse effect, sea level rise, flood, cyclone, storm and droughts have impelled the people to migrate as refugees."[24]

While explaining the distinction between migrants and refugees many experts emphasize the importance of both push and pull factors in the decisions of potential flight of people. Push factors are defined as characteristics of the current place of residence, such as low economic development that encourage people to leave. Pull factors are characteristics of a potential home that attracts people to go there. While drawing the line between refugees and migrants, the experts say, refugees are motivated mainly by push factors, migrants are motivated by a combination of both push and pull factors. According to Astri Suhrke, the environmental problems are more likely to produce migrants than refugees, because such problems usually develop gradually, which means that the push effect is not sharp and sudden and that pull factors can therefore clearly enter into potential migrants' calculation.[25]

While doubting the push/pull distinction as a helpful analytical tool, Homer Dixon instead notes that, "the real motivator of migration is the gap between the potential migrants' current level of satisfaction and the level they expect to attain in a new land. The larger the gap, the greater the incentive to migrate. And this gap is not determined by an additive function of push and pull factors but by the relationship between the perceived quality of life in the home region and that in the receiving region."[26]

The Bangladesh – Northeast India region is a good illustration of land scarcity having caused millions of people to move from Bangladesh to the Indian states of Assam, Tripura, West Bengal and Orissa and other metropolitan cities. According to estimates, migrants from Bangladesh have expanded the population of neighbouring areas of India by 12 to 17 milllion.[27]

Domestic Cleavages and Social Segmentation

Economic decline and population movement induced by the environmental scarcity of resources can create many fissures and cleavages in the political and social structure of the states. Severe environmental scarcity can aggravate divisions or segmentation among ethnic, religious, and linguistic groups.[28] Homer Dixon has cited the example of South Africa to explain how environmental scarcity can create division and segmentation within a society. According to him," scarcity sharpens distinction between winners and losers- between groups that profit from scarcity and those that are hurt. It encourages competitions among groups for control of resources critical to survival and prosperity, and it encourages resource-dependent groups to turn inward and to focus on narrow survival strategies.'[29] This insularity reduces interactions among these groups and between them and the state. Apartheid in SouthAfrica concentrated millions of blacks in some of the country's least productive and most ecologically sensitive territories where population densities were further increased by high natural birth rates. These black populations in these homelands were forced to subsist on local cropland, water and fuel wood. Lack of resource management skill, high population growth and low capital investment further eroded the available resource base. The corrupt local elites captured the best local resources throwing the weaker groups to the most ecologically sensitive parts of the homelands. As a result, wide areas of these territories were completely stripped of trees for fuel wood, grazed down to bare dirt and eroded of topsoil. A 1980 report concluded that narely 50 percent of Ciske's land was moderately or severely eroded, and nearly 40 percent of its pasturage was over grazed.[30] With the tumbling down of apartheid's strict urban migration laws, environmental scarcity in the homelands contributed to an annual migration of up to 750,000 blacks to the country's urban areas.[31] Sever population pressures in these areas damaged and depleted local streams, community cropland and fuel wood stocks. The concentration of many people on a degraded resource base in the context of weak local governments, again caused resource capture : powerful warlords secured community support in part by controlling basic environmental resources, such as land and water through their own paramilitary forces.[32]

Division and segmentation took place among the ethnic groups, among family based clans, and among residents of township, informal settlements, and work hostels as communities withdrew into themselves to protect access to these resources. As resources were degraded within their territories, warlords often tried to maintain power by pointing to resources in neighbouring townships and informal settlements and mobilizing their communities to seize them. This mobilization often set in motion a cycle of appalling violence among African communities. "Squatters are mobilized to fight for access to resources in neighbouring townships, and township youth organize military style units to defend their areas and counter-attack squatter areas."[33] These processes exacerbated inter group segmentation, broke down social networks and weakened community norms resulting in a loss of the society's social capital.

Similarly in Bangladesh, land scarcity induced population growth and environmental degradation and its consequences in the form of flood, cyclone and storms resulted in settlement of Bangladeshis in Chittagong Hill Tracts. This brought Chakmas, the original inhabitants of CHT into direct conflict with Bangladeshis, as the demands on the resource base increased[34] and the Bangladeshis occupied the traditional resource base of the Chakmas forcing them to environmentally marginalised slope hill areas.

Disrupted institutions and growing weakening of the state

The social effects described above are likely to tear the fabric of legitimized, accepted and authoritative institutions that guide and pattern social behaviour. [35] Homer Dixon and his group have identified the links between rising environmental scarcity and declining state capacity.

First, environmental scarcities increase financial and political demands on the state. For example, as they have cited China, Pakistan, SouthAfrica and Philippines, environmental scarcities put pressure on the countries to meet the needs of the expanding marginal people. The Chinese government must spend huge sums on new infrastructure such as wells, dams, canals, pipelines and irrigation system.[36] The study of the impact of environmental scarcities on state capability in these states shows that "environmental scarcities expand marginal groups that need help from government by constraining rural economic development and by encouraging people to move into cities where they demand food, shelter, transport, energy, and job."[37] As J.D. Kasarda and A.M.Parnell notes rapid population growth in cities contribute to 'high rates of unemployment and under employment,... soaring urban poverty, insufficient shelter, inadequate sanitation, inadequate or contaminated water supplies, serious air pollution and other forms of environmental degradation, congested streets, overloaded public transportation systems, and municipal budget crises."[38] In response the subsidies of urban services introduced by governments eat away their revenues causing misallocation and distortion of prices and capital. Second, resource scarcities affect the state via their effects on elites. On one hand, scarcities threaten the incomes of elites that depend on resource extraction. They often compete among themselves for cornering the shrinking resource rents and try to obstruct any institutional reforms that would distribute more fairly the costs of rising scarcity. As seen in China, water scarcities have engendered sharp divisions between Beijing and the provinces and among the provinces themselves.[39] Philip Shenon of the New York Times notes that critical land scarcities have emboldened these powerful groups to ignore the instructions of the state. According to him, "huge profits from land sales... tend to overwhelm any fear of the central government. With the prospect of millions of dollars in profits on even a small plot of land, developers and corrupt officials are often willing to take the risk of ignoring instructions from Beijing."[40] They often lobby to change the property rights and other laws governing the use of scarce resources such as water, land and forests. In many societies these rent seeking elites influence the state through bribery, kickbacks and other forms of corruption.[41]

Third, in such a situation of resource scarcity and the elites trying to corner the resource pie, struggle for resource control takes place between the powerful elite group and the weak groups. This infighting and struggle for shrinking resources worsens social segmentation and weaknes the cohesive fabric of civil societies. State's failure to meet the needs of the marginal people weakens its legitimacy. Fourth and finally, if resource scarcity affects the economy's general productivity, tax revenues to local and national governments can decline. Such a revenue decline hurts elites that benefit from state largesse and reduces the state's capacity to meet the increasing demands of the people arising from environmental scarcity. Charles Barber shows that the proliferation of local conflicts over forests depress the legitimacy, coherence and reach of the Indonesian state at the local level.[42]

Lack of Ingenuity Supply and Adaptative Failure

According to Homer Dixon and his group the environmental scarcity of resources and its various social effects can make the developing societies vulnerable to the extent that they would not be able to adapt to the situation and find out solutions as they are plagued by lack of supply of ingenuity. What adds to their disadvantage is the very fact that many of them are not adequately endowed with the social institutions including the productive research centers, efficient markets and capable states which are necessary for an ample supply of both social and technical solution to environmental scarcity. As argued by this group, environmental scarcity sometimes diminishes these countries' ability to create and maintain such institutions.[43] But the proponents of new economic growth theory or endogenous growth theory, overlooking the fact that environmental scarcity can stymie the supply of ingenuity, argue that ideas are a factor of economic production in addition to capital, labour, and land.[44] In their views scarcities will boost the supply of ingenuity. In other words necessity will often be the mother of invention. This will help alleviate scarcity's severity and social impacts. These economists focus mainly on ideas embodied in new technologies. Homer Dixon's concept of supply of ingenuity is not confined to mere technological and scientific innovations. That environmental scarcity and its social impacts need only the attention of scientists and engineers has been reviewed. According to Homer Dixon, "the supply of this technical ingenuity depends on an adequate supply of social ingenuity at many levels of society. Social ingenuity is key to the creation, reform and maintenance of public and semipublic goods such as markets, funding agencies, educational and research organizations, and effective government."[45] Social ingenuity is thus a precursor to technical ingenuity and key to adaptation strategies. Countries afflicted by environmental scarcity of resources need sophisticated and stable systems of markets, legal regimes, financial agencies, and educational and research institutions to promote the development and distribution of new grains adapted for dry climate and eroded soils, of alternative cooking technologies to compensate for the loss of firewood and of water conservation technologies. In the words of Dixon, "it is both an input to and output of the economic system."[46] It is not that the ingenuity required to adapt to the crisis created by environmental scarcity is produced always at the top of the social hierarchy, it is also produced at the community and household levels.

Burgeoning population growth and increasing per capita resource consumption are causing a steady increase in the ratio of the consumption of resource per unit of time to the total amount of resource available. Thus the scarcity has in turn increased the requirement for ingenuity. The increased pace of resource depletion means that societies have less time to adjust their institutions and technologies. Since human beings are consuming a very broad spectrum of renewable resources at very high rates, serious scarcities can affect many resources simultaneously or in quick succession. This multiplicity of scarcities can make it harder to find relatively abundant resources to substitute for scarce ones. Further, the ingenuity requirement to compensate for scarcities of renewables is generally greater than that of non renewables. The difference between the two has been ignored by the economists. There can be no substitute for renewables. For example, nothing can substitute for water in the tissues of a living organism. Thus, the options for responding to scarcity of renewables are restricted and society's need for ingenuity rises. Even if the requisite is supplied, the systems will still be vulnerable to sudden shocks, because many environmental scarcities are interconnected and unpredictable. Further more, even if societies manage their resource systems exceptionally well, fundamental physical, biological, and social constraints may make it difficult to fully compensate for the effects of scarcity. These constraints include inescapable time lags that affect a society's ability to mitigate scarcities, like the time needed for a new forest to grow, for a new technology to be disseminated, or for cultural change to take place.[47]

Constraints on Supply of Ingenuity

But optimists like Aaron Wildavsky, Todd La Porte, Dennis Avery, Thomas Schelling ensconce faith in the ingenuity and resourcefulness of man to find solution to the crisis. Julian Simon has abiding faith in the resoursefulness and capability of human beings to respond to the scarcity of rersources and the concomitant consequences with new ingenuity that will keep them invulnerable to the crisis and leave them better off[48] Homer Dixon does not harbour a slightest doubt of the extraordinary potential of human resourcefulness and enterprise, but nonetheless argue that some poor developing societies will not be able to supply the unprecedented amounts of ingenuity required to solve the impending environmental scarcity. He has identified four factors that can limit the supply of social and technical ingenuity: market failure, social friction, shortage of capital, and constraints on science.[49]

Market Failure

Many resources especially renewables like hydrological cycles and productive seas, cannot be physically controlled or divided into saleable units for the exclusive use of specific consumers. It is therefore hard to assign clear property rights and they remain open access resources that are vulnerable to over exploitation. As Repetto notes, "prices can guide market adjustment processes only if resources can be bought and sold which presupposes that they can be owned. Many important resources are not privately owned, usually because it is impossible to segregate one person's portion from another's and reserve it for exclusive use."[50] At best their scarcity is indirectly reflected in the prices of marketable resources that are dependent on them. For example, damage to ocean ecosystems can cause fish prices to rise.

Even if property rights are clear, market prices may still not fully reflect the costs of resource use. Resource extraction or use can produce negative externalities- such as river siltation from upstream deforestation- that are not incorporated in the resource price. The World Bank notes that "clear property rights may induce private users to adopt the correct pattern of timber extraction, taking into consideration the current and future benefits from logging, but give no incentive to take into account the costs of deforestation to those living outside the forests- for example, increased soil erosion or lost biodiversity."[51] Poor countries are particularly disadvantaged because they start with underdeveloped economic institutions. The state which is to play a key role in increasing ingenuity supply through research, development, and extension services is very much stymied in developing societies because of economic and other constraints.

Social Friction

As analysed above, environmental scarcity of resources produces frictions, cleavages and social segmentation among various groups. Mancur Olson's work [52] is cited by Dixon to understand his analysis. First, Olson shows that small coalitions can generally organize themselves more quickly and purse their interests with greater force than larger groups. They therefore have political power disproportionate to their size, and they can be more nimble, focused and effective in their lobbying. Second, this proportionate power will be particularly acute in unstable societies, because large coalitions need time and social stability to establish themselves and grow. The governments of unstable countries are therefore " systematically influenced by the interests, pleas and pressure " of small coalitions.[53] Third, small coalitions are always guided by their narrowly defined self interests in total disregard of the common well being. Olson writes, there is "no constraint on the social cost such an organization will find it expedient to impose on the society in the course of obtaining a larger share of the social output for itself."[54]

It has been pointed out earlier that increased environmental scarcity of resources exacerbated social segmentation, sharpens cleavages among groups, and provokes action by groups to protect their interests in terms of resource capture and rent seeking. Building on Olson's argument, Homer Dixon points out that the smaller groups pursue their interests better than larger ones in developing countries, because their possession of wealth, power and status well informed by their position in the social order had made this minority well advantaged while pursuing their narrow self interests, the small coalitions often impede the process of institutional building that reflect the broader interest of society. They hamstring all efforts to reform existing or establish new social institutions, laws and structural arrangements, if these effects run counter to coalitions interests, as they often do.[55] This social friction arising out of environmental scarcity makes it harder to focus and coordinate social activities, talents, and resource in response to scarcity. As Olson points out, narrow coalitions "interfere with an economy's capacity to adapt to change and to generate new innovations."[56] Thus, the coalitions provoked to action by scarcities will sometimes block solutions to the very same scarcities. Increasing social friction affects incentives for political entrepreneurs to produce and deliver the social ingenuity needed for new institutions, Ruttan and Hayani opine that "the supply of institutional innovations depends critically on the power structure or balance among vested interest groups in the society."[57] As seen, scarcities of cropland and water in Bangladesh – scarcities arising from both skewed resource distribution and populations growth – induce some agricultural innovation, but they also inflame existing distribution struggles among coalitions. These struggles in turn have of water control institutions and the deployment of irrigation technologies.[58] A society which is very much informed by a culture of selfishness will soon slip into narrow coalitions as scarcity worsens. For example, Filipino culture encourages cooperation within groups rather than among groups resulting in what is often remarked, clannishness of the society and undermining the concept of national welfare.[59] To the contrary, a society where there is prevalence of culture of good will, reciprocity, civic mindedness, and trust can impede the rise of narrow coalitions as scarcity worsens.

Thus Homer Dixon cited these two contextual factors which will especially influence the degree of social friction caused by environmental scarcity and the extent to which this friction hamstrings the supply of ingenuity. In his words, "if narrow coalitions have already penetrated the state, social friction will have a particularly strong effect on society's capacity to reform and build institutions." [60]

Capital Availability

Capital both financial and human is necessary to respond to the environmental scarcity of resources. Availability of capital is generally lower in developing societies. To undertake any scientific and technological research in response to scarcity needs huge finance. Political entrepreneurs need financial capital to provide selective incentives and side payments to coalitions that block institutional change. The state needs capital to provide public goods like infrastructure and resource monitoring. Many societies afflicted by severe resource scarcity are poor; moreover, the often predatory behaviour of their elites further reduces general capital availability.[61] Dearth of adequate and required laboratory equipments, computers, journals and books impedes the efforts to build indigenous research capability in developing societies.[62] In addition to shortage of finance, lack of human capital also cripple the supply of ingenuity. For example, the United Nations reports that by 1987 nearly a third of Africa's skilled labour had left for Europe and that the continents as a whole lost sixty thousand middle and high level managers between 1985 and 1990.[63]

Constraints on Science

As said above, the economic and technological optimists exudes faith in human kind's ability to manage the resource scarcity. Homer Dixon outlines four constraints on science that will hinder ingenuity's supply.[64] First are human cognitive limits. Humans do not have the infinite ability to understand and manage the complex, multivariate, nonlinear and often chaotic processes of ecological –social systems. The relationship in some of these systems are "simply too numerous and complex to be grasped , much less controlled, by the human intellect."[65] Many natural, social systems are probably computationally intractable, which means they cannot be accurately modelled using any realistic amount of computational power.[66] Scientists are additionally handicapped because they cannot conduct controlled experiments with large scale natural and social systems.

The cognitive features of human brain evolved in ecological circumstances restrict human ability to supply ingenuity. Second is the escalating cost of research in the field of science and technology. To probe deeper into the labyrinthine nature has become a very costly affair. The poor societies plagued by resource scarcity need the help of advanced science like molecular biology which they cannot afford, especially when faced with resource crunch. A third constraint arises from the cumulative nature of scientific knowledge : each new discovery must build on a host of earlier ones. "The pace of discovery is marked by jumps and lags as scientists make breakthroughs or lose time pursing fruitless leads. This pace cannot be easily forced, especially in basic science where the work's ultimate practical use is not clear."[67]

A final constraint is science's vulnerability to social control. It needs a great variety and abundance of resource inputs, a non hierarchical institutional structure, a dense network of connections between like minded innovators , and a popular culture for promotion of science. For example, the development in former Soviet Union in 1990s crippled its vast research establishments and led to the rise in occult and anti science movements causing a decline in respect for analytical thought.[68]

Conclusion

Many analysts and economists place undue faith in the infinite capability of human kind to produce ingenuity to deal with the environmental scarcity . So far as the developing societies are concerned, the supply of ingenuity will not be plenty because of factors and constraints analyzed above by Dixon and his group. There may be a mismatch between the severity of the crisis and the amount of ingenuity needed to adapt to the crisis. This ingenuity gap shall be a common phenomenon of developing societies where the scarcity of resources has been severe. This is the reason why the developing societies are more vulnerable to environmental scarcity than the developed world. The question is not about human adaptation to the environmental scarcity and its social effects whether by suffering, starving and bearing the amount of hardship. The fulcrum of the analysis lies on how to prevent the crisis. This is what comes within the broader ambit of environmental security. It is not that the environmental scarcity shall be allowed to bust into a crisis until a solution is to be found out putting boundless faith in the potential of human ingenuity, as the economists and critics suggest. Human ingenuity is not like a magic wand that unearths a solution out of its sleeve every time when we are spurred by necessity. The crisis shall not be kept on waiting till scientific investments and knowledge come out with a solution. The impact of social effects of environmental scarcity on future generations may have to be more severe and far reaching. Since the poor societies are afflicted by severe environmental scarcity, there will be more ingenuity gap and adaptive failure because of the constraints cited above. In that case the very onus and functions of maintaining intergenerational equity and responsibility may not be properly discharged. The scarcity if let go to be fully blown into a crisis, just keeping faith in the ingenuity of man to reveal a solution, the prospects for social friction and violent conflict not only in the present generation but also in future societies looms large. This analysis is not that of an apologist to put forth arguments linking environmental scarcity and conflict, or it is not that of an economic optimist to wait and see keeping undue faith in human ingenuity which is not plentily available to the poor societies. It is about defining environmental security that includes a pro active preventive measure to avoid the prospects for conflicts as outlined by the dominant environmental conflict paradigm, and efforts to redefine the conventional wisdom about the security based on realist frame work, and the western pattern of development based on fossil fuel based industrialization taking into account the intergenerational equity and intergenerational responsibility.

References

1. Thomas F. Homer Dixon, Environment, Scarcity and Violence,(Princeton, New Jersey : Princeton University Press,1999), p. 80, and Volker Boge,

"Analytical Framework to grasp environmental conflict." ENCOP, Centre for Security Studies and Conflict Research and Swiss Peace Foundation, Zurich/Bern, Occasional Paper no.I, July 1992, p. 4.

2. Vaclav Smil, Energy, Food, Environment : Realities, Myths,Options (Oxford : Oxford University Press, 1987).

3. Fakhri A. Bazzaz and Eric D. Fajer, "Plant Life in a CO_2 Rich World", Scientific American, Vol-266, no.I January 1992, pp. 68-74 ; Harold A. Mooney and George W. Koch, "The Impact of Rising CO_2 Concentrations on the Terrestrial Biosphere", Ambio Vol. 23, no. I, 1994, pp. 74-76.

4. Per Pinstrup- Anderson, World Food Trends and Future Food Security, Food Policy Report (Washington, D.C. : The International Food Policy Research Institute, 1994, p. 3.

5. Celestine Bohlen, "Conference on Food Aid Starts in Rome," New York Times, 13 November 1996, p. A5.

6. Thomas F. Homer Dixon , Environment, Scarcity and Violence (Princeton, New Jersey : Princeton University Press, 1999), p. 84.

7. Down To Earth, Vol. 9 No. 6, August 15, 2000, pp. 30-40 : See also Peter Hazell and C. Ramasamy , The Green Revolution Reconsided : The Impact of High Yielding Rice Varieties in South India (Baltimore : Johns Hopkins University Press, 1991).

8. Pierre Crosson and Jock Anderson, Resources and Global Food Prospects : Supply and Demand for Cereals to 2030. World Bank Technical Paper Number 184 (Washington, D.C. : World Bank, 1992), p. 12.

9. WRI, World Resource 1996-97, p. 238.

10. Mark Rosegrant, Mercedita Agcaoili- Sombilla, and Nicostrato Perez, Global Food Projections to 2020 : Implications for Investment, IFPRI Food, Agriculture , and the Environment Discussion Paper 5 (Washington, D.C. : International Food Policy Research Institute, 1995).

11. Homer Dixon , Environment, Scarcity and Violence , *op.cit.*, p. 82.

12. Volker Boge, Proposal for an Analytical Framework to Grasp Environmental Conflict, p. 4.

13. J.D.Longstreth et al., " Changes in Ultraviolet Radiation Reaching the Earth's Surface," Ambio Vol. 24, no. 3, 1995, pp. 153-65 : Harry Slaper et al., "Estimates of Ozone Depletion and Skin Cancer Incidence to Examine the Vienna Convention Achievements," Nature Vol. 384, no. 6606, 21 November 1996, pp. 256-68.

14. Per Pinstrup- Andersen, Global Perspectives for Food Production and Consumption," Tidsskrift for Land Okonomi vol. 4, no. 179, 1992, pp.145-69.

15. Jeffrey Leonard, " Overview", in Environment and the Poor : Development Strategies for a Common Agenda (New Brunswick, N.J. : Transaction, 1989), pp. 27.

16. Robert Repetto, "Balance Sheet Erosion How to Account for the Loss of Natural Resources", International Environmental Affairs Vol.1, no.2, 1989, p. 129.

17. *Ibid.*, pp. 103-37.

18. Kimberley Kelly and Thomas Homer Dixon, Environmental Scarcity and Violent Conflict : The Case of Gaza, Occasional Paper of the Project on Environment, Population, and Security (Washington, D.C. : American Association for the Advancement of Science and the University of Toronto, 1996).

19. Vaclav Smil and Mao Yushi, The Economic Cost of China's Environmental Degradation, Occasional Paper of the Project on Environmental Scarcities, State Capacity, and Civil Violence (Cambridge, Mass : American Academy of Arts and Science and the University of Toronto, 1998).

20. See Jodi Jacobson, Environmental Refugees : A Yardstick of Habitability, Worldwatch Paper 86 (Washington D.C. : Worldwatch Institute,1988); Arthur H.Westing, "Environmental Refugees : A Growing Category of Displaced Persons," Environmental Conservation 19, no. 3, 1992, pp. 201-7; Norman Myers, Ultimate Security : The Environmental Basis of Political Stability (New York : W.W. Norton, 1993); Norman Myers, "Environmental Refugees in a Global Warmed World", Bioscience 43, no.II 1993, pp.752-61; and Norman Myers and J.Kent, Environmental Exodus : An Emergent Crisis in the Global Arena (Washington D.C. : Climate Institute, 1995).

21. E.El Hinnawi, Environmental Refugees, UN Environmental Programme, Nairobi, 1985, p. 4.

22. Anthony H. Richmond, "Reactive Migration : Sociological Perspectives on Refugees Movements", Journal of Refugee Studies Vol. 6,1993, p. 6.

23. A. Giddens, The Consequences of Modernity (Cambridge : Polity Press, 1991), p. 171.

24. V.Suhrke and A. Vinsentin, "The Environmental Refugees : A New Approach", Ecodecision, September 1991, pp. 37-74.

25. Astri Suhrke, Pressure Points : Environmental Degradation, Migration, and Conflict, Occasional Paper No.3, Project on Environmental Change and Acute Conflict (Cambridge, Mass. : American Academy of Arts and Sciences and the University of Toronto, 1993); and Astri Suhrke, "Environmental Degradation and Population Flows", Journal of International Affairs Vol.47, no.2, 1994, pp. 473-524.

26. Thomas F. Homer Dixon, Environment, Scarcity and Violence, *op.cit*, pp. 93-94.

27. Homer Dixon, Environment, Scarcity and Violence, p. 95 ; Sanjoy Hazarika, Bangladesh and Assam : Land Pressures, Migration, and Ethnic Conflict, Occasional Paper No. 3, Project on Environmental Change and Acute Conflict,(Cambridge, Mass. : American Academy of Arts and Science and the University of Toronto, 1993), pp. 52-54 and N.Gaan, Environmental Degradation and Conflict : The Case of Bangladesh – India (New Delhi : South Asian Publishers, 1998), pp. 62-63.

28. Valerie Percival and Thomas Homer Dixon, "Environmental Scarcity and Violent Conflict : The Case of South Africa," Journal of Peace Research 35, no. 3, May 1998, pp. 279-98.

29. Homer Dixon, Environment, Scarcity and Violence, *op.cit.*, p. 96.

30. Francis Wilson and Mamphela Ramphele, Uprooting Poverty : The South African Challenge (New York – Norton, 1989).

31. Lesley Lawson, "The Ghetto and the Greenbelt", in Jacklyn Cock and Eddie Koch, ed., Going Green : People,Politics, and the Environment in South Africa (Capetown : Oxford University Press,1991), p. 47.

32. Anthony Minnaar, "Undisputed Kings : Warlordism in Natal", in Anthony Minnaar ed., Patterns of Violence: Case Studies of Conflict in Natal (Pretoria: Human Science Research Council, 1992), p. 92.

33. Doug Hindson, Mark Byerley, and Mike Morris, "From Violence to Reconstruction: The Making, Disintegration, and Remaking of an Apartheid City", Antipode 26, no.4, 1994, p. 341.

34. N.Gaan, Environmental Degradation and Conflict, *op.cit.*

35. Douglass North, Institutions, Institutional Change and Economic Performance (Cambridge : Cambridge University Press, 1990), p. 3.

36. Elizabeth Economy, Reforms and Resources : The Implications for State Capacity in the People's Republic of China, Occasional Paper of the Project

on Environmental Scarcities, State Capacity, and Civil Violence (Cambridge, Mass. : American Academy of Arts and Sciences and the University of Toronto, 1997).

37. Thomas Homer Dixon, Environment , Scarcity and Violence p. 101.

38. J.D. Kasarda and A.M. Parnell, eds., Third World Cities : Problems, Policies and Prospects (London : Sage, 1993).

39. Elizabeth Economy, Reforms and Resources, *op.cit.*

40. Philip "Shenon Gincum Journal : Good Earth is Squandered. Who will Feed China ?" New York Times, 21 September 1994, p. A10.

41. See Charles Barber, Indonesia Case Study, Occasional Paper of the Project on Environmental Scarcities , State Capacity and Civil Violence (Cambridge, Mass : American Academy of Arts and Science, and the University or Toronto, 1997).

42. Barber, Indonesia Case Study, *op.cit.*

43. Homer Dixon , Environment, Scarcity , and Violence, p. 108.

44. See Paul Romer, "The Origins of Endogenous Growth," Journal of Economic Perspective, vol.8 , no.1, 1994, pp. 3-22.

45. Homer Dixon, Environment, Scarcity and Violence, p. 110.

46. *Ibid.*

47. *Ibid*, p. 114.

48. Julian Simon, The Ultimate Resources (Princeton, N.J. : Princeton University Press, 1981) p. 345.

49. Homer Dixon, Environment, Scarcity and Violence, *op.cit.* p. 117-118.

50. Robert Repetto, "Population, Researces, Environment : An Uncertain Future", Population Bulletin, Vol. 42, no. 2, 1987, pp. 6-7.

51. World Bank, World Development Report 1992 (New York : Oxford University Press, 1992), p. 70.

52. Mancur Olson, The Rise and Decline of Nations : Economic Growth, Stagflation, and Social Rigidities (New Haven, Conn. : Yale University Press, 1982).

53. *Ibid.*, p. 165.

54. *Ibid.*, p. 44.

55. Elinor Ostrom,Larry Schroeder, Susan Wynne, "Institutional Incentives and Sustainable Development," World Development Vol.17, no.9, 1989, pp. 1375-87 ; Also see David Reed's review of the World Bank's World Development Report 1992 in International Environmental Affairs Vol.4,no.4, 1992.pp. 367-71.

56. Olson, Rise and Decline, p. 62.

57. Vernon Ruttan and Yujiro Hayami, "Towards a Theory of Induced Institutional Innovation," The Journal of Development Studies Vol .20, no.4, 1984. p. 213.

58. James Boyce Agrarian Impasse in Bengal : Institutional Constraints to Technological Change (Oxford : Oxford University Press, 1987) see also Homer Dixon, Environment, Scarcity and Violence,Chapter-3.

59. Richard Kessler, Rebellion and Repression in the Philippines (New Haven, Conn. : Yale University Press, 1989), p. 18.

60. Thomas Homer Dixon, Environment, Scarcity and Violence, p. 120.

61. Peter Evans, "The State as Problem and Solution : Predation , Embedded Autonomy, and Structural Change," in Stephan Haggard and Robert Kaufman (Princeton, N.J. : Princeton University Press, 1992), p. 144.

62. See Mamman Aminu Ibrahim, "The Information Drain : Obstacles to Research in Africa", Gatekeeper Series no.32 (London: International Institute for Environment and Development, 1992).

63. United Nations Development Program, Human development Report 1992, (New York : Oxford University Press, 1992), p. 57.

64. Homer Dixon , Environment, Scarcity and Violence, pp. 123-24.

65. E.O. Wilson, "Is humanity Suicidal ? New York Times Magazine, 30 May 1993, p. 29; Michael Manning and David Rejeski," Sustainable Development and Risk : A Fit ? Paper presented at the Conference on Comparative Risk and Priority Setting of Air Pollution Issues, Keystone, Colo., 7-11 June 1993.

66. See Joseph Traub and Henryk Wozniakowski,"Breaking Intractability", Scientific American Vol. 270, no.1, January 1994, pp. 102-7.

67. Thomas Homer Dixon, Environment, Scarcity and Violence, p. 124.

68. Robert Wright, "The Experiment That Failed : Why Soviet Science Collapsed", The New Republic 205, no.18, 28 October 1991, pp. 20-25; and Sergei Kapitza," Anti Science Trends in the USSR", Scientific American 265, no. 2, August 1999, pp. 32-38.

6

Violent Environmental Conflict

The environmental scarcity of resources and its main negative social effects contribute to violent conflict. If there is a decline in agricultural productivity, decline in economic wellbeing, poverty stalks their way of life in developing societies, if a large number of people move from their homelands, if societies segment and fractured and institution disrupted and state weakened, what types of conflict are likely to occur ? A host of theories on the nature and types of conflict can be dependent upon to answer this question. Three sets of theories in particular are key : one each at the individual, group, and systemic level of analysis.

Based on the study of individual psychological behaviour frustration-aggression theories are generally used to explain civil strife, revolution, insurgencies, strikes, riots and coups. These theories propound that people get aggressive when they feel frustrated at not getting their things or wishes fulfilled or being hindered by some one from fulfilling a strong desire. An important subset of these theories suggests that this frustration and aggression can be caused by relative deprivation, which means that when people perceive a widening gap between the level of satisfaction they have achieved and the level they believe they deserve. Deprivation is therefore relative to some subjective judgement of equity or fairness, and the size of the perceived gap obviously depends on the beliefs about economic justice held by individuals.[1]

Using social psychological theories of group behaviour, group-identity theories explain inter group conflicts on the basis of nationalism, ethnicity, and religions. These theories explain on what basis the individual members identify themselves with a group ultimately to be bound by itself distinguishing it from other groups and resulting in a we vs. they syndrome. Identifying with a group bestows on them a psychological sense of camaraderie or we ness which is reinforced and satisfied in a group when it is directed against and attacks another group. Similarly a person's sense of self worth may be strengthened when his or her group's status is enhanced in relation to that of other groups. By attacking or denigrating outside or the other groups the leaders of a group exploit the psychological need of its members for feeling of we ness, to enhance their political power and pelf within their own groups.[2] Structural theories premised on the assumption of micro economics and game theory, explain conflicts arising from the rational calculations of actors in the face of perceived external constraints. The structure of an actor's social situation is the perceived set of possible interactions with other actors and the perceived likely outcomes of these interactions. This structure is determined by physical factors such as the number of actors in the system, resource limits, and barriers to movement or communication; by social factors such as shared beliefs and understandings, rules of social interactions, and the set of power relations among actors in the system; and by psychological factors, in particular, the beliefs and preferences of other actors.[3] As Wendt notes, the view of structure as constraint is only one of three possible positions on the issue. Structure can also be thought of as generating the actors rather than constraining them, or structure and actor can be seen as dialectically related- that is, generating but not reducible to each other as Gidden suggests.[4]

Divided into two parts general and civic strife – the general structural theories are often applied to inter state war. These theories suggest that external constraints such as power imbalances in the international system can encourage or compel countries to go to war. Scholars who emphasize such constraints usually also acknowledge the causal importance of internal factors, such as the actor's particular interests and beliefs. By stressing power relations almost exclusively perhaps Waltz comes closest to presenting a purely structural theory of international behaviour and war. De Mesquita contends that the geographic proximity of actors is an important structural determinant of international conflict, where as Choucri and North emphasize differences in states' resource endowments.[5] Structural theories of civic strife suggest that revolution, insurgency, rioting, and coups de'tat are a function of the opportunity structure that confronts groups challenging the authority of the state. In the words of

Homer Dixon," serious civil strife is not likely unless the structure of political opportunities facing challenger groups keeps them from expressing their grievances effectively and peacefully, but offers them openings for violence against authority."[6] This opportunity structure depends on the relative power and resource of challenger groups or the state, on the power of groups that might ally themselves with challenger groups or the state, and on the costs and benefits that groups believe they will accrue through different kinds of collective action in support of or in opposition to the state.[7] On the basis of these theories, Homer Dixon and his group have identified three main kinds of conflict that might arise from environmental scarcity: simple scarcity conflicts, group identity conflicts, and insurgencies.

Simple Scarcity Conflict

Here the use of the adjective ' simple' doesn't mean 'unimportant'. It distinguishes this type of conflict from others that involve psychological and social processes more complex than those posited by rational – choice theorists. Volker Boge terms these conflicts in a narrower sense that emanates directly from the overuse and/ or pollution/destruction of a renewable natural resource.[8] For example : If a downstream riparian threatens to go to war against an upstream riparian, because the latter pollutes river water so gravely that it cannot be used by the inhabitants of the downstream riparian, who are highly dependent on this water, then the 'environmental' character of conflict is evident. Four environmental resources such as agriculturally productive land, forests, river water and fish would appear to ignite simple- scarcity conflict. Evidences can be culled from history showing the number of wars among or between nations over non-renewable resources. For instance, prior to and during second world war, Japan was bent on securing coal, oil, minerals, and other resources in China and South East Asia. Hitler's force before capturing Caucasian oil fields were stopped at Stalingrad in 1942. The Gulf wars in 1991 and in 2003 launched by the U.S. and its allied forces against Iraq were at least partly motivated by the desire for oil.[9] The lateral pressure theory as argued by Nazli Choucri and Robert North suggests that countries facing high resource demands and limited resource availability within their territories would seek the needed resources through trade or conquest beyond their boundaries.[10] It has been the conventional wisdom based on the realist paradigm of international relations theory, that critical scarcities of natural resources can cause war. Realism focuses on states as rational maximizers of power in an anarchic international system. State behaviour including conflictual behaviour is mainly a function of the structure of power relations among states in the system, not of events or pressures operating inside these states. Because environmental scarcity's most important social effects unfold inside countries, realism therefore encourages scholars to ignore or downplay the consequences of environmental problems for peace and conflict. If these scholars do consider environmental problems, they focus disproportionately on simple scarcity conflicts among states over strategic environmental resources such as water.[11]

While accounting for causes of international conflicts, these scholars have not been able to distinguish between the scarcities of renewable and non-renewable resources. They have overlooked two reasons why modern states do not generally fight over renewable resources. If power is understood in a dispositional sense, then renewable resources such as cropland, forests and fish seized from a state cannot be easily translated into power. These resources may eventually generate wealth that can be harnessed by the state for its own ends, this outcome is uncertain and remote in time. In contrast, states can quickly use non-renewable like oil and iron to build and fuel the military machines of national aggression. The more easily fungibility of non renewable resources into state power than the renewable resources in a political and economic order based on fossil fuel industrialization and militarisation, has enhanced the chances of interstate wars over the non-renewable resources. That does not in any way denude the renewable resources of their potential to state power and for violent conflict among nations. In the seventeenth through nineteenth centuries, for example, shortage of timber for naval ships contributed to serious, and sometimes violent conflict among European powers.[12]

The second reason is that the countries with agrarian economies highly dependent on renewables tend to be poor and poor countries cannot easily buy large and sophisticated weapons to attack their neighbours. According to Homer Dixon , "for these reasons, both the incentives and the means to launch resource wars are likely to be lower for renewables than for non-renewables." [13]

Water Conflict

The exception is water particularly river water. Water is a critical resource for personal and national survival. It is essential to the production and use of military power, and rich countries are as dependent on water as poor countries.[14] About 40 percent of the world's population lives in the 220 river basins shared by more than one country. At a meeting in Stockholm in August, 1995 Ismail Serageldin, the World Bank's Vice President for Environmentally Sustainable Development, declared that the 'wars of the next century will be over water not oil.[15] The same words were echoed by Wally N Dow, Secretary General of the United Nations Conference on Human Settlements, held in 1996 in Istanbul when he stated : "I suspect that in the next 50 years, we will see a shift from oil to water as the cause of great conflicts between nations and people".[16] In a similar vein, Peter Gleick stated: "As we move into the twenty first century, water and water supply systems are increasingly likely to be both instruments of political conflict and the objectives of military action as human populations grow, standards of living improve, and global climate changes make water supply and demand more problematic and uncertain."[17] The likelihood of water being the object of military or political actions depend on the degree of water scarcity; the extent to which the supply is shared by two or more groups ; the relative power of these groups ; and the ease of access to alternative sources.

Water Conflict in Middle East

Water is unevenly distributed throughout the world. Some regions receive enormous amounts of rainfall or river flow, while others are extremely dry and suffer from absolute scarcity. Human factors, such as high population densities or intensive industrial development, may cause conditions of relative scarcity. In the past, absolute scarcity was a more important factor in regional water issue. Increasingly, however, relative scarcity is the problem leading to water disputes, as population pressures come up against resource limits.[18] Water has been used as a military and political goal not only in the past but also in the modern era. The Jordan River Basin has been the scene of a wide variety of water disputes among Israel, Syria, Egypt and Lebanon. The Jordan river is formed by the convergence of three tributaries- the Hasbani, Dan, and Banjas that rise in Lebanon, Israel, and Syria respectively. The three meet in Israel territory and flow southward into the sea of Galilee (Lake Tiberias) and then into the Lower Jordan River. The Yarmouk tributary which rises in Syria forming boundary between Syria and Jordan, and then between Jordan and Israel enters the Lower Jordan a few kilometers South of Lake Tiberias. The lower Jordan which flows into the Dead sea, forms the boundary between Israel and Jordan for some of its distance. In the late 1950s when Israeli government began work on its National Water Carrier to take fresh water from Lake Tiberias and transport it Southward, outside the Jordan basin and throughout Israel, and announced to complete its first phase by 1964, the Arab countries strongly protested as violation of their security.[19] In response to Israel's diversion plan, the Arab League adopted a proposal to divert the Hasbani River into Lebanon, channel its surplus waters to the Banias in Syria, and divert the surplus waters of the Banias to the Yarmouk for the benefit of Jordan. In 1964, within weeks of the work on the Arab diversion project, a series of military responses to rival water projects involving Israel, Syria, Jordan and Egypt took place. It became a cardinal political issue of interstate relations in the Middle East. Israeli's Premier Levi Eshkol stated that "water is a question of life for Israel," hence any attempt to prevent Israel from using a portion of the Jordan waters would be considered a violation of its right to exist. He warned the Arabs that "Israel would act to ensure that the waters continue to flow" into its territory.[20] Border clashes, and arms race between Israel and its Arab neighbours continued throughout 1966 till all these resulted in Arab- Israeli war of 1967. The Jordan water crisis of 1964 and Arab- Israeli war of 1967 were two distinct crises in a protracted conflict . In the analysis of Lowi, in the years leading up to war, water related issues-specially, the depletion, degradation or altered allocation of water resources, or the threat there of – were definitely part of the larger picture" forming " part of the package of issues that established conditions that made war likely."[21] The 1967 war in the Middle East resulted in Israel winning control of all the headwaters of the Jordan as the ground water of the West Bank. Water remains an important factor in the politics of the region. The multilateral and bilateral peace talks conducted in the 1990s, which led to the interim agreement between the Israelis and the Palestinians and to the Peace Treaty between Israel and Jordan explicitly included negotiations and agreements on the shared water resources of the Jordan river. Israeli and Syrian concerns over the Banias, which originates in the Golan Heights, remain an important unresolved issue. Jordanian concerns about Syrian dams on the Yarmouk, the major tributary to the Jordan, are still unanswered.

The use of water as an instrument or tool of conflict and as both offensive and defensive weapons which has a long history, can be seen at present. For example, North Korea announced plans in 1986 to build a major hydroelectric dam on the Han river upstream of South Korea's capital, Seoul. The project would provide much needed electricity to the North, but South Korea perceives its potential as a military weapon. In the perceptions of South Korean hydrologists, the destruction of the dam by the North would destroy most of Seoul. In response to this South Korea has built a series of levees and check dams above Seoul to defend against any such threat.[22]

During the Persian Gulf war in 1991, the coalition forces discussed the possibility of using the Ataturk dam in Turkey on the Euphrates river to shut off the flow of water to Iraq solely dependent on flows of the Euphrates river.[23] Both Syria and Iraq continue to have an ongoing dispute with Turkey over the operation of Ataturk and the level and quality of flows in the Euphrates reaching both downstream countries.

Water Conflict in South East

Singapore not self sufficient in water because of its high population density and small size, depends on piped water from Malaysia for nearly half of all its needs. In addition, Singapore imports water from Malaysia that it then treats and sells back under an agreement signed in 1965.[24] Their relations have been strained by economic competition, religious, political and ethnic differences. In early 1997 these relations soured again when ruling political party urged its government to review the basis of its water agreement with Singapore.[25] Singapore clearly worried that Malaysia might use water as a political and strategic weapon against Singapore. In response to this Singapore has launched also a campaign to increase water and reduce consumption through an aggressive conservation programme which includes new desalination plants that would produce water at about eight times the cost of current supplies.[26] Whether this exorbitant cost is worth risking the strained relations with Malaysia, remains to be seen.

Where water resources and water supply systems are a strategic resource with economic, political or military importance, they become targets during war or conflicts. The United States targeted irrigation levees in North Vietnam. Syria tried to destroy Israel's National Water Carrier while it was under construction in the 1950s. The Iraqis in1991 Gulf war intentionally destroyed the water desalination plants of Kuwait and in turn suffered from the destruction of their water supply system by the allied forces assembled to liberate Kuwait.[27]Some other important water disputes can be cited.

The Nile River

The Nile is critical to Egypt, which gets no usable rain and lacks other sources of water but has a few rapidly diminishing aquifers under the desert. Egypt's 65 million people escalating to 75 million in a decade are solely dependent on the river. The Nile flows through eight other countries before reaching Egypt. According to a 1959 agreement with Sudan, the Egyptians are entitled to 55 billion cubic meters of Nile water a year, while Sudan gets 18.5 billion. Total water availability in Egypt was 63.5 billions cubic meters in 1990, but in 1998, demand for water already had scored to 68 billion and it has been rising rapidly since. US AID office predicted in the mid 1990s that Egypt could experience a 16-30 per cent water deficit by the end of the century.

Egypt threatened in the past that it was willing to go to war to prevent any one upstream from tampering with its water flow. In the mid 1980s, it was on the point of ordering air attacks against Khartoum for perceived water threats. What has saved Egypt until now is that both Sudan and Ethiopia were too poor and wrecked by civil war to contemplate any large-scale water diversion. The current tensions between Egypt, Ethiopia and Sudan are in a way a continuation of two thousand years old struggle over who will control the waters of Nile.[28]

In 1993, a general agreement was reached between the new Ethiopian government and Egypt pleading that each country would not do anything to the Nile that might harm the other. But it is not so much what the Ethiopians have done, as Jan Hultin said in a survey for Leif Ohlsson's Hydropolitics but rather what they might be doing. Former UN Secretary General Boutros Boutros Ghali, when he was Egypt's Foreign Minister in1990, pointed out that the national security of Egypt, which is based on the water of Nile, is in the hands of other African countries. Later, he was even more blunt," the next war in our region will be over the waters of the Nile", echoing a comment made by Anwar Sadat fifteen years earlier.[29] With Egypt adding another million people every nine months, demand is already in critical conflict with supply. In 1998 border clashes between Egypt and Sudan and clashes between Addis Ababa and the Eritreans took place exacerbating the conflict already ingrained in them. The current anxieties have only sharpened the potential for conflict. It is not just Egypt whose demand is increasing. Development, that Holy Grail pursed by all nations will inevitably up the demand all along the river. The supply, however cannot be changed.

Euphrates River Conflict : Turkey, Syria and Iraq

Both the Tigris and the Euphrates rise in the moisture rich and green mountain valleys of eastern Anatolia in Turkey. The Tigris flowing Southeast across the low mountain valleys and Turkish plains marks briefly the border between Syria and Turkey. Then it heads Southeast through Iraq, passing Baghdad and the Central Marshes before joining with the Euphrates to become the Shatt–al–Arab, which flows into the Persian Gulf. The Euphrates starts nearer the Black Sea and moves through a series of Turkish lakes and reservoirs before crossing into Syria and its Euphrates Dam. From there it heads in a Southeasterly direction, crosses into Iraq and traverses the parched desert South of Baghdad before flowing through the Hammar Marsh and so into the Tigris not far above the Iraqi town of Basra.

The Euphrates and its tributaries form Syria's major water resource on which hinges the country's need for increasing food production to match its burgeoning population. Iraq downstream from both Turkey and Syria is solely dependent on both rivers. Nearly 85 per cent of Iraq's population gets all their water needs from the twin rivers. The claims and counter claims of the three countries are in many ways mutually incompatible, complicated by ethnic conflicts and historical memories. Turkey, as the upstream country maintains it has absolute sovereignty over any water originating in its territory. " The water is as much ours as Iraq's oil is Iraq's comes close to outlining Turkey's policy.[30] Iraq downstream argues its case on the rivers' natural course and on its historical rights to waters enjoyed by the people of the Southern Mesopotamia since the dawn of civilization six millennium ago. Midstream

Syria argues with forked tongue citing the prior-use doctrine to Turkey on the one hand, and the sovereignty doctrine to Iraq on the other. With water scarcity getting worsening day by day, the Turkish Kurds inhabiting headwaters of both rivers and fighting for independence using guerrilla actions against hydro installations as a weapon of policy, here exacerbated this built-in rivalry.

Though Turkey is water rich, it does not have much water to spare. In 1995 divided among the country's 60 million citizens, its total water supply which stands at 195 cubic kilometers annually, gives some 3.174 cubic meters for person per year. What alarms Turkey is its growing population which is projected to reach 78 million by 2025, and some put it closer to 90 million. As assessed by hydrologists, the country's available water which stands at 95 billion cubic meters annually, would reduce the per capita availability to around 1800 cubic meters, though adequate but not plenty. What has stirred the hornet's nest is Turkish $32 billion Great Anatolia Project (GAP) comprising twenty two dams, a network of irrigation canals, weirs and barrages on the Euphrates. This project which has originally scheduled to be completely by the late 1990s, was put off until 2010. This will reduce 40 per cent of Euphrates water to Syria, from 32 billion cubic meters to 20 billion.[31] Syria is already plagued by short of water with an annual water availability of not much more than a thousand cubic meters per capita.[32] The Euphrates meets 80 to 85 per cent of the water needs of its towns, industries and farms. Syria has been vulnerable to chronic drought in recent years. The water that passes through Turkey and on to Syria is laden with agrochemical, fertilisers, pesticides and salts posing serious threats to the cropland and agricultural productivity. With population growing at more than 3.5 per cent a year – one of the highests in the world- is in danger of running out of water by century's end, even without GAP. With Turkey's going ahead with the GAP, the crisis will certainly culminate in a catastrophe. According to Mikhail Wakil of the University of Ale Ppo, "within the next two decades Syria will experience an ever increasing water deficit that will have serious effects on its social and economic development... A water crisis is emerging throughout the Middle East, creating the potential for future conflicts".[33]

Iraq for its part would lose somewhere between 80 and 90 per cent of its Euphrates allotment. Iraq still has the Tigris, but that too, will be affected by projected impoundments. Neither conservation nor more intensive use of smaller rivers would be able to make up the shortfall. Even without the GAP, consumption targets for the three countries exceed the total supply by 18.3 billion cubic meters per year on the Euphrates system dam, and 5.8 billion for the Tigris. Both Turkey and Syria have locked themselves in an exchange of acrimonious words over the situation. Syria is too weak relative to Turkey to rectify the situation by force. Instead, it has provided sanctuary to guerrillas of the Kurdish Workers Party (PKK) which has long been launching an insurgency against the Turkish Government in eastern Anatolia. Turkey fears that Syria might use its moral and material support to the guerrillas as a leverage in bargaining over the Euphrates water. In October 1989, the then Prime Minister, Turgut Ozal suggested that Turkey might impound the river's water if Syria did not restrain the PKK.[34] On the other hand, Iraq has a stronger position vis-à-vis Turkey in view of the fact that it is not entirely dependent on the water of the Euphrates as it has the Tigris virtually all to itself. Another reason is that it is able to coalesce the flow of water from Turkey to the flow of oil into Turkey. Because of its own high level of water consumption and the history of hostilities between the two nations, Syria could be unwilling to join hands with Turkey against the downstream riparian Iraq. Similarly, a joint approach by Syria and Iraq is inconceivable, governments in the region link solution to the Kurdish problem directly and indirectly to water issue. According to a report, "there is very little likelihood, in the present at least, that the water problems in the Tigris, -Euphrates basin will escalate into armed conflict over the distribution of water resources. However, if the riparians continue to insist on unilaterally defined consumption targets, the security risks will mount... Indeed, it is precisely the radical transformation of the basin's environment, mediated through the socio-economic problems of the rural producers in all the three countries that may destabilize international security."[35]

The Orontes river Conflict : Turkey and Syria

The Orontes river which rises in Lebanon and flows through Syria into the region of Hatay, is the cause of yet another bone of contention between Syria and Turkey. For years, Syria successfully bullied Lebanon into not damming the Orontes, wanting most of the water for itself. The river is effectively dry by the time it reaches Turkey, with a paltry flow of 25 million cubic meters out of the potential total of 1.2 billion, and most of that has already been used by Syria. The solution has been elusive as Syria evades any talk with Turkey. Any agreement would imply Syrian recognition that Turkey owns Hatay. On the other hand Turkey is very much alive to Syria's hypocrisy – Syria and Iraq both argue against Turkish appropriation of Euphrates on the ground of material harm to the downstream countries". But here, where Syria is the upstream country, it blithely does what it would not allow Turkey to do elsewhere, and takes all the water for itself.".[36]

Yarmuk River Dispute : Israel and Jordan

The Jordan basin (The Jordan and Yarmuk rivers) is the country's sole source of running surface water. In the 1950s, Jordan tried to gain control of its water resources by building storage system on the Yarmuk. Israelis destroyed them in1967 war and have on repeated occasions since, prevented maintenance being done on the Jordanian dams and canals. Withdrawal of water was seen as threats to Israeli farmers in the Jordan Valley. In the 1970s, Jordan came up with a plan to build a dam at Makarin on the Yarmuk, near the border with Syria. Israel objected, under World Bank financing rules, Israel had to be consulted. It was too large and it would take too much water from the system.

The growing relevance of the waters in the Jordan basin can be attributed above all to population growth in the region, the over exploitation of water resources especially through the explosive growth of irrigated agriculture and water pollution. The waters of the Jordan and Yarmuk are crucial to the existence of Israel and its Arab neighbours. The rights to water courses have played a decisive role since the Israeli state was first established. Johnston's Unified Plan of 1955 containing proposals for sharing the waters of the Jordan and Yarmuk among Israel, Jordan, Lebanon, Syria and the West Bank failed to be implemented due to the lack of consensus from Syria and Lebanon. Israel maintained that its quota should be larger than originally allocated, because its territory had increased to Golan Heights. Jordan, for its part did not see why unilateral annexation of the Golan should oblige Israel with larger share of water than allocated to it.[37] Taking advantage of this unregulated situation, Israel was drawing almost twice as much water from the Jordan in the early 1990s as envisaged by the water sharing regime in the Johnston's Unified

Plan.[38] The dispute between the Israel and its neighbours is closely linked to the question whether the allocations according to Johnston's Unified plan have effectively been shifted in Israel's favour as a result of later's territorial gains," In the current situation, the conflict over use and allocation of the water resources must be seen as a conflict of interest as opposed to a conflict over values, with water being assessed absolutely and not relatively".[39] In 1999,Jordan applied to the World Bank for money to build another dam called Wahda on the upper Yarmuk with Israel again objecting on the ground that this would affect its ability to cope with ever escalating domestic water demands. King Hussain was driven to say that water is the only reason that could bring Jordan to war with Israel. On October 24, 1994 Israeli – Jordan Peace Treaty was signed containing detailed provisions as to quantitative allocations, protection of water quality and the requirement to manage and develop their water resources without causing harm to the other party. An unusual feature of this treaty is the recognition that existing resources are not sufficient to meet their demands and that the signatory parties must therefore cooperate in order to develop new water resources. Expanding Jordan's supply of water could only be achieved otherwise if Israel were to limit its own share – a concession that would be unacceptable to the Israeli population. Very little progress in implementing the treaty commitments has been achieved to date.

Lure of the Litani river : Israel and Lebanon

The Litani is one of Lebanon's major rivers, with a flow of about 580 million cubic meters a year, smaller than the Jordan. The Israelis have never really given up thinking of the Litani as rightfully theirs. It is difficult to know how much water motivated Israel to invade Lebanon in 1978 and again in 1982 and to maintain its security zone there. Some analysts suggested that Israel government was motivated by "hydraulic imperative" to attack Lebanon to seize control of the Litani, a wholly national river that is located, at its Southern most point, within a few miles of the border with Israel.[40] Another view is that there is no credible evidence to suggest that the "lure of Litani", drew Israel into a long and costly war with Lebanon."[41] Though the aim was the need to smash PLO guerilla and terrorist bases in Lebanon, but water undoubtedly played a significant role.[42] It is said that during and after war, Israel had never made any efforts to divert a portion of the Litani waters into its territory. But a Food and Agricultural report in the mid 1970s based on a slender evidence, declared that Israel had built underground pumps along the Lebanese border to suck water southwards. In 1984 the Lebanese government lodged a complaint to the United Nations about Israel's supposed diversions, not only of the Litani but of two small rivers that conveniently fell into the security zone controlled by Israel.[43] By virtue of having set up a security zone in the South of Lebanon, Israel assumed control over headwaters of the Hasbani tributary. Hence by the military operation of 1982, Israel could become an upstream riparian in the Jordan basin, as it assumed control over all three tributaries of the main trunk of the Jordan system. Whether this was the main motivation behind Israel's invasion, remained unproven, but "that it was an outcome is certain".[44]

Conflict in Golan Heights and West Bank: Israel and Palestine

The depletion of overstretched ground water has reached such a critical point that it has become a bone of conflict between Israel and West Bank. What adds to the exacerbation of the crisis is that approximately 40% of Israel's sustainable annual supply of ground water, or one quarter of its total renewable freshwater supply, originate in Palestinian territory that from 1967 till this day has been under the occupation of Israel. As said, by the mid 1990s Israel's overexploitation of its aquifers stood at 2100 million cubic meters, against a supply that ranges in good years from 1950 million cubic meters to 1600 in drought years. Worse than that Jordan's drawing was 20 per cent more water that it received. As a result, the coastal aquifers in the region, especially in the critical Gaza area have been overdrawn out leading to huge seawater intrusion which has become a major problem. Adding to this is the political discrimination exercised against the Palestinians or Arabs who are not allowed to dig their wells deeper than those by the Israelis. According to Israeli's own figures, by 2010, the water deficit in Israel would reach 360 million cubic meters, Jordan's 200 million and West Bank's 140 million. In such a critical situation of environmental scarcity of water, the conflict between Israel and West Bank is likely to blow into a war like situation.

These waters flow naturally across the Greenline- the Armistice Demarcation Line established in 1949-into Israel, where they are tapped and integrated into Israel's National Water Carrier System. Since two thirds of the water Israel uses originates in territory it now controls through military conquest in the West Bank and Golan Heights, and its overall dependence on these waters, stringent measures have been adopted by Israel's Water Commission Administration to ensure that no more than 20% of the ground water go to the Palestinian population and the rest 80% fully assured to the Israelis across the Green Line and in settlements on the West Bank.[45]

What is perceived as a greater security threat to Israel is not the territorical claims of Palestinians on West Bank but the already overdrawn environmental water scarcity which could be further worsened in case West Bank goes to Palestinian control. This explains the reason, "why Israeli is not willing to relinquish its suzerainty over West Bank. As a result, the Israelis could be denied access to the portion of water supply it is till date dependent on, and intensive pumping of aquifers within the territory could degrade the quality of ground water that flows into Israel beyond the Greenline.

Control over this national resources is considered equally important for the Palestinians, as the water resources are key to the socio-economic-development of a future Palestininan state. This explains clearly the intimate linkage between national security concerns and environmental scarcity of resources. For this reason in the peace talks that were initiated in 1991 discussion of water issues was put off until final negotitation.[46] That does not in any way diminish the centrality of water issues to the future negotiation between Israel and Palestine.

The Okavango River Conflict

The Okavango river is shared by Angola, Namibia, and Botswana and is the largest river in Southern Africa. Most of the flow originates in Angola, flows southeast to Namibia and then turns south into Botswana. The river drains into the Okavango delta- a world renowned ecosystem and a major tourist spot. Several years ago Botswana proposed a major project called the Southern Okavango Integrated Water Development Project in order to provide water for irrigation, urban uses, livestock and a large mine. International Union for the Conservation of Nature (IUCN) was critical about its environmental implications. Though the project is of much interest to Botswana, it is put on hold due to above environmental concerns. More recently, the relationship between Namibia and Botswana has been bedeviled by Namibian plans to construct a 250 kilometers pipe line to divert water from the Okavango River to eastern Namibia and its capital of Windhoek. Further it intends to build an emergency pipeline to connect their Eastern National Water Carrier with Okavango to deal with severe drought and water scarcity. While an Okanvango River Basin Commission (OKACOM) comprised of Angola, Botswana, and Namibia was formed in September 1994, there is no long-term agreement over management or allocation of the river and no agreement that this project should go forward.[47] Negotiation and discussion between Namibia and Botswana are continuing. Some reports have come up suggesting that Botswana is increasing the size of its defense forces in response to this dispute, and there is concern in South Africa about the potential for conflict between the two countries over this issue.[48] According to Peter Mokaba, Deputy Minister for Environmental Affairs in South Africa, a change in the balance of power in the region could lead to an arms race.[49]

Rivers Of Kruger National Park : Tributaries to the Limpopo and Incomati Rivers

Six major rivers flow east from the Republic of South Africa into Kruger National Park and then into Mozambique, where they ultimately join the Limpopo or Incomati rivers. All these rivers and tributaries are highly threatened by burgeoning population and utilization. Kruger is the most important tourist center in South Africa region and also the most important centers of biodiversity. According to Peter Van Niekerk, "SouthAfrica is very aware that the Mozambicans are unhappy about the quantity of water reaching them. In fact, the Mozambicans were so unhappy that they declined to sign a proposed Memorandum of Understanding".[50] Mozambique has even threatened to move the World Court against South Africa. This was a move to put pressure on South Africa to address Mozambique's concerns which would tarnish South Africa' image both within the subcontinent and in the Organization of African Unity. The end of the war in Mozambique and the considerable improvement in the relation between South Africa and Mozambique opens up new possibilities that water issue can be addressed explicitly and directly.

Water Scarcity : Pakistan Vs. India

The Indus and its five tributaries – Jhelum, Chenab, Rabi, Beas and Sutlej constitute the largest irrigation system in the world. Independence and partition in1947 divided the Indus water system between India and Pakistan giving rise to concerns in Pakistan that its water needs would be curtailed by upstream development activities in India.

Indo-Pak Water Conflict

In exercise of her sovereign right to exploit the water vitally needed for her existence and development, India in 1948 cut off water supplies to West Pakistan by diverting water from Sutlej, Beas and Rabi Rivers in the Punjab, a border state divided between India and Pakistan. For five weeks, 1.5 million acres of land in Pakistan received no water. This action on the part of India threatened the Pakistan agricultural output during the growing spring season when the water demand was high. This led one authority to describe the situation," No army with bombs and shell fire could devastate a land so thoroughly as Pakistan could be devastated by simple expedient of India's permanently shutting off the sources of water that keep fields and people of Pakistan alive".[51] Water appeared to be a potent weapon in the hands of India to wreak disaster on Pakistan's agriculture.[52] The dispute was eventually resolved through the World Bank intervention, and Indus Water Treaty was signed finally in 1960 under which India was allowed exclusively rights to the waters of Ravi, Sutlej and Beas rivers. The waters of the western rivers, Indus, Jhelum and Chenab were assigned to Pakistan except for limited specified uses in occupied Kashmir and India. The Punjab water issue has been one of a series of problems contributing to the 1965 Indo-Pakistan war and to the unabated arms race in the subcontinent.[53] Punjab, Rajasthan and Haryana and New Delhi, states of India and Pakistan are highly dependent on the Beas, Sutlej and Ravi rivers for irrigation and power.

The Indus water treaty was planned not on the basis of interdependence but on the basis of respect for division of sovereignty between the two adversaries. Without showing concern for the environmental impact of the diversion two independent river systems were created. Environmental degradation especially increased salinity and bacterial life in the water and surrounding soils has affected the productivity of agriculture and the quality of drinking water.[54] For Pakistan, however, its inferior power resources, downstream riparian position and acute dependence on the Indus waters made it highly vulnerable to its adversary upstream. In the context of history of bitter relation between India and Pakistan on the issue of Kashmir, the eventual degradation of the Indus Waters attendant to hydro engineering alteration of the river system did not loom larger than the immediate concern for independent territorial and political sovereignty on the basis of physical division.

Tulbul Nagivation Project Dispute

The dispute is over the proposed construction by India of the Tulbul Navigation project which Pakistan calls Wular Barrage on the Jhelum downstream for the Wular lake. India wants to construct this barrage to control the flow of water in the Jhelum in the lean season October to February- the flow of water in the Jhelum is about 2,000 cubic feet per second and its depth is about 2.5 feet. However, year round navigability requires the flow and depth of water to be almost double of this. The barrage would make the river navigable between Sopore and Baramula. Although work on the barrage started in1984, it was stopped after Pakistan protested in1987.

Pakistan's contention is that building a barrage on the Jhelum at the mouth of Wular amounts to storage of water in contravention of the Indus Water Treaty. According to the treaty, India was allowed the limited use of the western rivers for domestic and agricultural purposes, run of the river hydroelectric generation and any non-consumptive use that would not diminish the flow of water to Pakistan. India is also allowed limited storage of water of the western rivers- a general storage capacity of 300,000 acre feet on the various channels of Jhelum (excluding Jhelum Main) and 10,000 acre feet on the Jhelum Main itself.

The basic controvery is whether the barrage on the Wular is a storage work or whether it amounts to controlling the water for navigation, a permissible activity under the Indus Water Treaty. The Indian stance has been that this is neither an act of storage nor of impounding the water of the Jhelum but of controlling the flow for navigation. The volume of water flowing to Pakistan would remain undiminished. The issue was taken up by India with Pakistan in1987. Since then eight rounds of secretary level talks were held to settle the issue bilaterally. In October 1991, the two sides came to an agreement.

The draft agreement mainly stipulated (a) that India would keep 6.2 meters of the barrage ungated with a crest level at EL 1574.90m; (b) that India would not make any alternation in the salient features of the project without mutual agreement between the two countries; (c) that India shall forego a general storage capacity of 300,000 acre feet out of the provision permitted to it on the Jhelum (excluding the Jhelum Main); (d) that in return for this the water level in the barrage will be allowed to attain the full operation level of 5177.90 feet- the timing of filling the lake will be decided by the two Indus Water Commissioners and should they fail to reach an agreement the filling of the lake would be between 21 June and 20 August; (e) that except for the stipulation regarding the filling of the lake, India would let all the waters entering the Wular lake downstream ; and (f) that all differences will be settled under the provisions of the Indus Water Treaty.

Later, however, Pakistan wanted that in addition to the above, India should give a commitment that it will not construct the 390 megawatt Kishenganga Hydroelectric Project as it would affect its proposed Neelum-Jhelum Power project. India has not given such a commitment. This was evident as the 1991 joint draft was reproduced verbatim in the non-paper on the subject handed over to Pakistan in January 1994. Thus is where matters stood since then.[55]

Ganges Water Dispute between India and Bangladesh

Water sharing of the Ganges is one of the most serious and disputed issues that has bedeviled the relation between India and Bangladesh. Conflict regarding the sharing of the water resources in the Ganges-Brahamputra river system between India and Bangladesh can be traced to 1947 partition of Indian sub continent into India and Pakistan . The Ganges which was solely India's before partition became an international river in the definition of Lemarquand with "a common property resources shared among the basin states".[56] The problem arose when India began to divert increasing volume of water for irrigation from the Ganges system, and construct a barrage across the Ganges near Farakka to solve the problem of the silting up of the Hooghly with a view to preserving the Calcutta port.[57] Bangladesh protested it on the ground that it could have harsh results for its downstream cropland, fisheries, and villages.[58] Though barrage's effects have contributed to the migrations out of Bangladesh into India, there is little chance of a water war here between upstream India and downstream Bangladesh. Bangladesh is so weak that the most it can do is to plead with India to release more water.[59] Despite many agreements and negotiation between the two countries, water issue remained unresolved until a final India-Bangladesh Treaty on sharing Ganges water was signed in1996 settling at rest the crisis for the present. The treaty, however has drawn considerable flak from different quarters, especially from Bihar, an Indian state. Bihar utilizes a large amount of water from upstream Ganga for irrigation purposes and its grouse is that the allocation to Bangladesh would adversely affect irrigation development in the state. If consideration of lean season requirements at Farakka remains uppermost while giving clearance to any irrigation project in the Ganga basin in the future, as assured by the then India Prime Minister Gowda to West Bengal Chief Minister, then it means cutting down any further utilization of the Ganga by the riparian states of Bihar and Uttar Pradesh. Thus, the problem of upstream use, leading to a lower quantum of water for Farakka, has for long remained unresolved.

With Sheikh Hasina out of power in Bangladesh , the Present Prime Minister Begum Zia has been providing safe sanctuary to Pakistan's ISI, Assam's ULFA terrorists and other insurgents to the chagrin of India . At present the Government of India asked its counterpart to stop helping insurgent activities against India from its soil. In case of water conflict in future, Bangladesh may use it as a pressure tactics upon India asking it to be soft towards Bangladesh on augmentation of flow of the Ganges water.

From the above analysis though one can glean that there has never been a single war fought over water, the potential for conflict remains always high. With water getting scarcer and scarcer day by day and degraded coupled with high population growth, the main crisis of the 21st century will be over water. Then above scenarios of conflict may turn out to be violent.

Environmental Scarcity of fish and Conflict

Another renewable resource besides water that has engendered conflict between nations is fish. Indo- Sri Lanka conflict over Kachchativu Island can be a best illustration of the linkage between environmental scarcity of fish and conflict. The decline in fish catch as noticed in South

Asia Seas Regions due to reasons mentioned in an earlier chapter recently can result into a conflict between India and Sri Lanka over fishing rights of local fishermen from Rameswaram, Nagapattinam and Padukottai in Tamil Nadu in the waters around Kachchativu, a tiny Sri Lankan island across the international boundary line. What lures the fishermen from areas of Indus to Sri Lankan waters beyond Kachchativu up to Delft Island off the Jaffna coast even at the risk of being shot by Sri Lankan navy is abundant prawns. The ocean currents and sedimentation on the Sri Lankan side of the Palk Strait have made it a rich field of Tiger prawns. Besides, while the fish in the Indian waters have been overharvested considerable lack of fishing activity on the Sri Lankan side after the outbreak of current ethnic crisis in 1983, has made the Sri Lankan waters fertile breeding grounds for prawns.

The genesis of the conflict between India and Sri Lanka is to be found in the maritime agreements signed on 26 June 1974 and23 March 1976 between them. The 1974 Agreement demarcated the boundary between India and Sri Lanka in the Palk Strait and ceded Kachchativu to Sri Lanka. The 1976 Agreement demarcated the boundary in the Gulf of Manner and Bay of Bengal and barred each country's fisherman from fishing in the other's waters. Both these agreements as subsequent events show, affected seriously the livelihood of Tamil Nadu fishermen.

Despite strong evidences on Indian side,[60] the Indian government ceded Kachchativu to Sri Lanka in 1974 ostensibly to mend fences and court friendship with its neighbours.[61] The loss of Kachchativu island was replenished by an explicit recognition in1974 Agreement of traditional rights of the Indian fishermen and of pilgrims to visit Kachchativu to take part in the St. Anthony's festival as they were used to doing in the past. Article 5 of the 1974 Agreement which purports to have protected the rights of Indian fishermen and pilgrims, says subject to the foregoing, Indian fishermen and pilgrims will enjoy access to visit Kachchativu as hitherto and will not be required by Sri Lanka to obtain travel documents, or visas for these purpose.[62] Further, Article 6 explicitly states, "The vessels of India and Sri Lanka will enjoy in each other's waters such as rights as they have traditionally enjoyed therein."[63] To display any doubts about Indian fishermen's traditional rights over fishing in Kachchativu, Swaran Singh, External Affairs Minister of India, while replying to

M. Kalyanasundaram, Member of Parliament, spelt out in candid words "the western side of the fishery line was exclusive fishery right of the Indian citizens and to the east of that was the right of the Sri Lanka fishermen. But in spite of that division," he said, "the fishermen were generally free to fish even roundabout Kachchativu and they also used Kachchativu island for drying their nets. About the traditional rights", he further clarified that "although Sri Lanka's claim to sovereignty over Kachchativu has been recognized the traditional rights of Indian fishermen and pilgrims to visit the island remain unaffected".[64]

The 1976 Agreement was entitled "Agreement between India and Sri Lanka on the Maritime Boundary between the Two Countries in the Gulf of Manner and the Bay of Bengal and Related Matters". There was an Exchange of Letters on the same day between India's Foreign Secretary Kewal Singh and Sri Lanka's Secretary to the Ministry of Defence and Foreign Affairs W.T. Jayasinghe, constituting an agreement between two countries.

The Indian Minister of External Affairs Y.B. Chavan stated in the Parliament while laying the text of the agreements, that "the two countries shall exercise full sovereignty and absolute jurisdiction on their side of the maritime boundary. Both countries have agreed that after the determination of the maritime boundary, fishing vessels and fishermen of one country shall not engage in fishing in waters of the other".[65]

The trouble started when Sri Lanka government held a different interpretation that the Exchange of Letters and the 1976 Agreement superseded Articles 5 and 6 of the 1974 Agreement and denied the Indian fishermen their traditional rights to fish in the Kachchativu waters. On the contrary, there is a definitive school of thought in India which finds no connection between the 1974 Agreement and the 1976 agreement and exchange of letters and used to read Articles 5 and 6 of 1974 agreement along with Swaran Singh's statement in the Parliament, so as to justify the continued traditional fishing rights of Indian people in the Kachchativu waters.[66] With the conflagration of ethnic crisis between LTTE and Sri Lanka, the conflict took a different turn more particularly after the assassination of Rajiv Gandhi on 21 May 1991, when Sri Lanka government tightened their security belt in order to block the movement of LTTE men between Tamil Nadu and Sri Lanka and stopped the Tamil Nadu fishermen from smuggling fuel, medicines and weapons to meet the needs of LTTE. With Sri Lanka imposing a " no-go-zone " in its northern territorial waters, since September 1993 the number of casualties by the Sri Lanka Navy firing on Indian fishing boats have gone up. As estimates show, by 1994, 84 of Tamil Nadu fishermen fell victim to the Sri Lanka Navy, 162 were injured, about 60 boats were destroyed and several hundred boats wrecked. The Tamil Nadu fishermen continued to be shelled by the Sri Lankan navy, as Tamil Nadu fisheries department officials admitted.[67]

In 1995, the Sri Lankan navy reportedly opened fire on 20 occasions on fishermen from Tamil Nadu, killing 20 and injuring 35. Twenty-five boats were damaged and two sunk. In January 1997, there were two incidents-one fisherman was killed and one injured. The second of these was on 20 January, when I. K. Gujral, the then External Affairs Minister of India was in Colombo, holding talks with Sri Lankan leaders. A concerned Chief Minister of Tamil Nadu, M. Karunanidhi, wanted Gujral to categorically tell Colombo that there should be no such incidents. He also warned of dire consequences in the state if the firing continued.

In 1991the fishermen of these regions of Tamil Nadu struck work and paralysed the local pisci centered economy of Tamil Nadu in protest against Sri Lanka government's ban on Indian fishing rights. Arulanandam, the Secretary of Rameswaram Verkottu fishermen's Association says, "losing Kachchativu waters to Sri Lanka is like losing a major limb of our body".[68] It affected livelihood of 25,000 fishermen's families. As prawn is not available in the shallow Indian coastal waters and has dwindled significantly due to degradation of the marine environment, nearly 15000 mechanical fishing trawlers used to stray into forbidden waters to bring in a total catch of 35-40 tonnes of prawns a day between the peak season September and February. This meant, an earning of Rs.100crores during this period was lost to Indian foreign exchange.[69] In response to the protestations in Tamil Nadu and the heavy damage done to it in terms of human lives and economy by Sri Lanka military forces, Jayalalitha, the Chief Minister of Tamil Nadu wrote a series of letters in September/October 1993 and February 1994 to P. V. Narsimha Rao, the Prime Minister of India urging upon him to issue a note of retaliation to Colombo. While in her 1991 Independence day speech she expressed her determination to fight with the centre to get back Kachchativu, in all these letters a mollified tone was found when she stressed on the need to restore the traditional rights of Indian fishermen as promised in1974 Agreement.

The issue figured in the talks between Prime Minister H.D. Deve Gowda and Sri Lanka President Chandrika Kumartunga in New Delhi in December 1996. It was again on the top of the agenda when I. K. Gujral met Sri Lankan Foreign Minister, Lakshman Kadirgamar in January 1997. But a solution eluded them.

In the face of an already dwindling fish catch in South Asia Seas Region[70] the question whether getting back Kachchativu is central to the protection to the Indian fishermen' rights assumes significance. Since ceding of Kachchativu has already been a fait accompli, the abrogation of 1974 Agreement would have serious ramification in view of India's maritime agreements with Maldives, Myanmar, Thailand and Indonesia. Loss of economic livelihood of Tamil Nadu's fishermen and the social effects triggered off by the dwindling fish catch in terms of economy, poverty and other health hazards due to absence of proper protein intake, will make it compulsive on the Tamil Nadu government to put pressure on the central government. It would want to restore the traditional fishing rights to the Tamil Nadu fishermen as enshrined in1974 Agreement. The talks in Colombo on 8 and 9 March between the two governments may not have been successful as expected.[71] This is an indication that the centre can no longer show a shortshift to a vital issue that affects miserably the very livelihood of thousands of fishermen in coastal Tamil Nadu .

Fisheries Department officials of Tamil Nadu suggested a charting agreement as a long term solution. According to this agreement, Kachchativu could be leased to India so that fishermen would have exclusive rights to fish there except in the security zone delineated by the Sri Lankan government. A short-term solution would be to cut a channel in the sand dunes of Adam's Bridge so that the mechanized boats could go down South to fish in the Gulf of Mannar. K. Marimuthu, a leader of the fishing community finds in the proposed Sethusamudram project a long-term solution. This project proposes to deepen a corridor within Palk Strait in order to facilitate the passage of ships, which are currently bound to touch Colombo port.

In case, the ethnic crisis in Sri Lanka comes to an end the Sri Lankan fishermen will be frequenting Kachchativu waters. Conceding to Indian demands for restoration of fishing rights by Sri Lanka will mean a cut into the economy and livelihood of their people. Prospects of required fish catch to meet the growing needs of the people appears to be bleak due to global warming and stratospheric ozone depletion which affects the sea life. As reports say, fish catch in Sri Lanka is declining due to its unregulated export of species of fish and change in marine water quality.[72] Kachchativu, which was once a sheer rock with no strategic significance to India, has in recent years developed into the making of a crisis between India and Sri Lanka.[73] The conflict between Tamil Nadu fishermen and Sri Lanka over fishing in Kachchativu waters has already been a conflict between India and Sri Lanka. Thus the group vs. state conflict has all the potentialities of a state vs. state conflict. In the absence of any amicable settlements between the two states keeping in view the economic needs of their people in the context of dwindling fish catch, the issue is likely to bedevil their bilateral relations.

Group – Identity Conflict

Here the ecological issues lead to conflict in a broader sense, i.e. the conflict results from the social effects of ecological degradation. Group identity theories described above explain the group identity conflict. In general group identity conflicts often emanate from the large-scale migration of population that can be caused by the environmental scarcity. As different ethnic, religious and cultural migrants meet an indigenous group due to environmental stress, an intergroup hostility and violent clash with a strong identity dynamic takes place. When the people started to move because their homeland has been severely damaged and depleted by deforestation, soil erosion, salinization etc, and came in contact with a different group, violent clash occurs which can be termed as an environmental conflict in a broader sense.

The situation in the Bangladesh – Assam region is a good illustration to show group- identity conflict.

Bangladesh represents a Malthusian nightmare with too "little food, or too many people on too little land". It is one of the world's most densely populated and impoverished states having 128 million people and GNP production per capita equivalent to 170 dollars. The country's population density of 930 per square kilometer is the world's highest. As studies show, nearly 56.5% of rural households comprising 7.8 million people without land and it is estimated that the proportion of landlessness has grown faster than the population.[75] In the words of development economist of Bangladesh "the economic viability of Bangladesh has long been in question because of its over population, poor natural resource base, vulnerability to natural disasters and undiversified economy dependent on the production of crops, rice and jute".[76]

Their economic deprivation and poverty is further compounded as seen before, by the region's vulnerability to frequent floods and cyclones. Bangladesh passes through recurring vicissitude of floods, cyclones, drought and famine destroying hundred and thousand human lives in the process. More than one million people have been killed by floods since 1961. The 1988 flood, the most devastating in terms of human lives and economy reduced rice production by 1.6 million tonnes and caused 1.3 billion dollars worth of damage to Bangladesh's economy.[77] Experts in Bangladesh opine that the single most harmful environmental factor in the country is flooding resulting in temporary and permanent migration not only across districts, but also across international frontiers.[78] Each year in Bangladesh 18 to 19 million people are affected by floods.[79] The displacees of the river bank erosion are the worst of the landless poor.[80]

The search for land due to increased landlessness caused by population growth and flooding has resulted in massive migration of landless to urban areas within the country and to India. It is seen in Bangladesh "land has always been almost as fluid as the water that shapes it. The geography of the delta changes like a speeded version of continental drift."[81] Every year new lands are formed due to river erosion by eating into the existing land. This is perceptible in Hatia and South of Noakhali. These lands are known as 'charland', the scramble for which has reached a crisis proportion in Bangladesh. People dispossessed by erosion have no automatic right to these newly accreted land. All these newly deposited charland belongs to the government. The landless as they are illiterate and ignorant and incapable of bureaucratic appeasement, are deprived of their lands with no access to this new land. Jotdars and rich landowners get hold of legal titles to charland under fictitious names, through their nexus with political and bureaucratic leadership. Apart from the woes which the poor landless peasants and fishermen suffered at the hands of landowners, jotdars, the absence of mangroves which are to stabilize their charland results in another human tragedy when tens of thousands of people are swept away by tropical storms sometimes with their islands disappearing into the sea.[82] This search for land has resulted in group clashes in Bangladesh with the state alternately supporting the landowners. This only reinforces the existing social divides and relative deprivation among the landless and poor people in rural and urban areas. The growing number of squatters around urban centres as in Bangladesh due to the migration of landless people has also resulted in clashes for jobs and economic sustenance.

Taking advantage of this feeling of relative deprivation among the people, military dictatorships frequently come to power. In Bangladesh frequent political instability punctuated by military dictatorships may be attributed to this growing sense of relative deprivation among the poor who are promised in return for their upliftment, votes. One of the primary reasons of the likelihood of democratic regimes being not successful in Bangladesh can be traced to the deepening poverty and internal calamities like floods, cyclones and surges. Poverty induced by environmental change leads to severe internal strike, including coup de tat; revolutions and insurgencies.[83] It is envisaged that these regimes that succeed to power through these means will probably be more extremists, authoritarian, corrupt and abusive of human rights.[84] In the absence of any concrete poverty alleviating measures to blot out relative deprivation existing among the poor upon which the ruling regimes cash in to come to power, could search other rhetorics to legitimize their continuity in power, and divert people's attention from the existing domestic failures. In Bangladesh, rising anti-India feelings of water issue or Babri Masjid demolition at Ayodhya in December 1992 or pandering to fundamentalist Muslims, have been observed in the recent years. This creates a feeling of continuous animosity towards the neighbouring states.

Induced by poverty and environmental degradation between 12 to 17 million landless poor people in their search for land and economic livelihood, have been migrating to neighbouring states of India, viz. Assam, West Bengal, Tripura, Orissa and urban centres like Delhi and Bombay creating a "sub nation, the size of Australia within India."[85] Though Bangladesh continues to proclaim that it has no illegal aliens in India,

Bangladesh communities have sprung up in Delhi and Bombay under the shelter of flyovers and in Muslim dominated regions. The political parties in India resort to political use of Bangladesh settlers as potential vote banks. While the Government of India put the Bangladeshi migrants figures at about 100,000, the intelligence sources estimate it at 200,000.[86] Providing advantage to political parties during hustings, these refugees receive patronage in terms of education, development and health facilities even at the cost of the local inhabitants. This contains the seeds of further local conflicts.

An estimate suggests that as many as 1,000 people a month are coming to West Bengal.[87] Most drift to Calcutta where their living conditions are marked by squalor, overcrowding and abysmal poverty. After the destruction of the Babri Mosque, many in the squatter communities were involved in the rioting and carnage that enveloped the city for several days.[88] It is seen that rioters were driven more by class hatred, envy and need to plunder and pillage the middle class than by any religious fanaticism.[89] Such rioting and clashes have domestic political fall out with some Hindu dominated political parties being inclined to reap political dividends. In their search for vote banks among the refugees, the political parties particularly the Congress and Communist Party have down-played the number of refugees settled in West Bengal.

Assam like situation in Orissa

The migration of Bangladeshis is not confined only to North eastern states of India but also has spread to Orissa's coastal areas. The ninety kilometre long coastal belt spreading from Dhamara of Bhadrak district to Paradeep of Jagatsingpur district has been littered with the illegal settlement of Bangladeshis including Mahakalapada, Dangamala and Ekakola of Kendrapara district. With the active connivance of local politicians, they have settled along these coastal areas destroying twenty five thousand hectare of reserved forest cover and converting these either to cultivable land or for prawn cultivation.

The destruction of mangrove forests for agricultural activities, unplanned and illegal habitation, and illegal prawn culture by these migrants could be one of the factors that helped super cyclone in Orissa in1999 grow in intensity and severity to surpass the previous cyclones. As Assam and Central governments are confronted with the problems of law and order, terrorism and insurgencies, the Orissa government is likely to face similarly these problems in near future. As reported, thousands of Bangladeshis are settling near the defence establishment at Bhadrak and understood to have been employed as daily wagers in the works of the defence institution in the night. These Bangladesh migrants are seen permanently settling in the nearby areas and getting BPL cards and other government facilities with the active support of politicians to turn them into a potential vote bank. In these coastal areas if reports of their being involved in narcotic traffics, ISI activity on behalf of Pakistan and other nefarious activities like smuggling, murder and burglary, are to be believed, then government's sitting silently putting the issue of illegal migration on the backburner has drawn broadside from the local people. This has irked the local people and their fear that this could convert the coastal belt of Orissa into a mini Bangladesh has resulted in strong resentment against the government failure to send back these migrants. A total of 1200 refugees of Bangladesh legally settled in1957 has now touched 1 lakh 80 thousand after 46 years causing a feeling of being reduced to minority in the local people.[90] One should not underestimate the fact that the settlement of Bangladeshis along the coastal line cutting deep into the forest cover - the natural bulwark against storm and cyclones - and the local people's economic livelihood, could one day turn into a violent conflagration between the migrants and local people with the considerable involvement of the Orissa government, like the violent one which happened recently in Raighar and Rangabhati of Nabarangapur district between the local tribal people and Bangladesh refugees settled in these areas by the central government over land. This has claimed the lives of 10 people, and injured 55 people. This is not a new one, adding to this, 56 times the local tribal people have been in skirmishes with the Bangladesh refugees. With strong pressure on the government not to show shortshrift to the migration of Bangladeshis, the government has decided to deport some Bangladeshis to their homeland.[91]

Due to the constant stream of migrants from Bangladesh to Assam, "the problem the Assmese face today is that it soon may be themselves who constitute an island within a sea of Bengalis".[92] The demographic structure of Assam has already been in a danger of having transformed the once dominant Assamese into a minority within Assam.[93]

There has been a spate of data rasing controversies about the exact number of illegal Bangladeshi migrants settling in Assam.[94] Estimates indicated that Assam's population increased from 14.6 million to 22.9 million during 1970s.This increase of 65.6% in one decade cannot be attributed to a natural population growth alone, especially when the rest of India, showed an increase of 24.7% during the same period.[95] Illegal migration is the only reason for that unexpected growth. Conservative estimates put it at over 6 million persons.[96]

The most recent study on this shows that the Indian government has been faltering in giving the correct estimate in order to reap political dividends by using them as legitimate voters. The Bangladesh government similarly for political reasons has been downplaying the number which may create conflict with India. Taking into account many expert estimates, this study is of the view that Bangladesh refugees have expanded the population of the neighbouring Indian states by 12 to 17 million of which 1 to 2 million may have been caused by the 1971 war. The population of Assam has gone up by at least 7 million people to its current total of 22 million.[97] Some accounts put the figure as high as 13 million and accuse the Bangladesh government of pursing a policy of lebensraum in north-east India.[98] The Assamese people however reluctantly accepted these migrants so long as the latter preferred to be mere labourers. Social tensions and violent conflict soon began to erupt when the migrants made serious inroads into their economy and put pressures on existing resource base. Along with this Assamese fear of being reduced to a minority within their own homeland leading to the extinction of their culture and identity, accentuated the problems.

The Assamese legitimate fear, discontentment and lowered standard of living was accomplished by the failure of both the central and state government's disinclination to stem the tide of flow of Bangladeshis who were settled by them for their votes.[99] This led to the first popular movement by All Assam Students Union (AASU) against these illegal immigration. Their main demands were mainly the detection of foreign nationals on the basis of the 1952 National Register of Citizens, their disenfranchisement and deportation. Strikes, demonstrations, blockades, civil disobedience and boycott of elections paralyzed the administration. A series of governments rose and fell, political instability continued to invite the intervention of central government and its rule. From 1981 to 1983, illegal immigrants swelled in numbers in Assam from Bangladesh resulting in a number of communal clashes and riots killing hundreds and thousands of people. The initial ethnic tension simmered for years due to the government's negligence and continued Bangladesh influx had suddenly turned into a fury of human carnage. It was the worst ethnic riots in India since independence, and the number of human lives lost in these riots surpassed even the anti-Sikh riots of 1984, or the anti-Muslim riots of Bombay in January 1993 or religious rioting in 1990 and 1992 over the Babri Mosque demolition.[100] The holding of elections in Assam in 1983 without fulfilling the AASU's demands resulted in extreme violence claiming lives of many thousands. A study of the reasons behind the riots underlines a common factor, i.e. land between Bangladesh and northeast India.[101]

The first step towards a solution to the ethnic conflict in Assam was the 1985 Assam Accord which outlined (i) the expulsion of all immigrants entering Assam after 24 March 1971, (ii) a ten year disenfranchisement of illegal immigrants entered between 1966 and 1971, (iii) and those who came before 1966 to be granted Indian citizenship.[102] In the general election Assam Gana Parishad created out of the AASU won the election ousting Congress party from power.[103] The Congress party ruling at the centre delayed in implementing the Accord which contributed to the continuance of the instability.[104] Renewed agitation by the AASU started again with United Liberation Front of Assam (ULFA) and Bodo Land tribal movements against the Bengalis joined in making revolutionary demands for balkanization of Assam.

The issue of migration of Bangladeshis has already soured the relations between India and Bangladesh. Erecting a fence along the parts of the border and the forced repatriation of refugees at any point on the border have contributed to continued conflict with Bangladesh. Despite a series of meetings at the ministerial and secretarial levels, illegal migration remains and continues unabated.

The ULFA and Bodoland movements continue in their demands for separatism are militarily encouraged by Pakistan and Bangladesh. As Hiteswar Saikia, the former Chief Minister of Assam revealed, the underground members of the United Liberation Front of Assam, the Bodo Security Force and other insurgent organization active in Manipur,

Nagaland and Tripura were getting shelter, weapons and arms training in Bangladesh and Tripura with the help of Pakistan's ISI.[105] The outgoing BSF director general, D.K. Arya of India in his 10 page press release statement disclosed that insurgent groups operating in North East were acquiring sophisticated weapons from Thailand with the active connivance of Bangladesh's intelligence agencies. Bangladesh is providing all possible help in terms of shelter and training and the ISI of Pakistan is also actively involved in encouraging the Indian insurgent groups operating from the Bangladesh soil.[106]

The press release further accused Bangladesh of fanning religious fundamentalism and disruptive forces and deliberately facilitating migration of its nationals to India. He further implied that rise of insurgency in North East and growing migration to India with Bangladesh and Pakistan actively involved in encouraging morally and materially such centrifugal forces, have serious implications on India's security and unity.

By the year 2020,there will be 220 million Bangladeshis and one billion Indians competing for land and jobs.[107] In the case of sea level rise, as predicated due to global warming, Bangladesh will be worst hit thus forcing some more million environmental refugees to enter India. This will compound the problems for India and will further downgrade Indo-Bangladesh relations.

If the influx is not stopped, there will be a sharp decline in communal harmony and centre-state relations will be vitiated giving rise to separatist tendencies and ethnic crisis in the north eastern part of India on a scale larger than what is seen at present. Any drastic measure to stop the flow of Bangladesh immigrants on the part of India will aggravate the conflict between India and Bangladesh. With the BJP party in India drawing attention to the issue of illegal Muslim migrants to enhance its sphere of influence especially in communist ruled West Bengal and north eastern states, the anti- India feelings of fundamentals Muslims in Bangladesh will be capitalized on by the ruling party in order to legitimize its authority. The BJP government's decision to complete fencing on the border has already raised a hornet's nest in Bangladesh.

Insurgencies

The third type of violent conflict in the form of rebellion or guerrilla war against state according to Dixon, can be explained and predicted by a combination of relative-deprivation theories and structural theories of civil strife.[108] These two theoretical perspectives in combination suggest that insurgency is a function of both the level of grievance motivating challenger groups and the opportunities available to these groups to act violently on their grievances.[109] Multiple pressures exerted at different levels in society interact to produce the grievance and opportunity simultaneously and raise the chance of insurgency. Environmental scarcity and degradation of resources plays a key role in exacerbating economic hardship and dislocation, intergroup segmentation and weakening institutions such as the state. These factors combine to cause insurgency. Economic decline- one of the social effects of environmental scarcity – as suggested by relative deprivation theory, will result in some groups being increasingly frustrated and aggrieved by the widening gap between their actual level of economic achievement and the level they feel they deserve. The faster the growth of the gap, the greater is the intensity of the grievance of the disadvantaged groups prompting it to act violently against those groups perceived to be the agent of their economic deprivation and misery.[110] There is no clear correlation between economic hardship and violence. Whether or not people get aggrieved and turn violent in the face of severe economic difficulty depends on their notion of economic justice. People belonging to a culture may adapt to situation of economic deprivation and unequal distribution of wealth and will not be prone to violence. Relative deprivation theory incorporates this contextual factor because deprivation is always relative to an individual or group's notion of economic justice. Economic crisis must be severe, persistent and pervasive enough to erode the legitimacy of the dominant social order and system of government. Group identity conflict and insurgency are the most important types of conflict likely to arise from severe environmental scarcity. It also contributes to several other types of conflict. Under certain circumstances, for example, scarcity can stimulate coup de'tat as aggrieved and conflicting elites take advantage of a weakened state. Drought and a severe food shortfall in Ethiopia opened up structural opportunities for the 1974 coup against Emperor Haile Selassie.[111]

In some cases environmental scarcity stimulates a form of insurgency that combine both group – identity and relative deprivation motivations. If a historically defined identity of a group is linked to a particular set of natural resources or a particular pattern of resource use, degradation of that resource can engender a feeling of relative deprivation. This relative deprivation may be expressed through aggressive assertion of group – identity. Sporadic violence in South Eastern Nigeria in1994 and 1995, when the Ogoni people protested pollution of the Niger delta by multinational oil companies, and the secessionist rebellions on the island of Bougainville in Papua New Guinea from the late 1980s to mid 1990s against a foreign owned copper mine which damaged the island's environment, are good examples to show the relative deprivation conflict.

Environmental Scarcity, Relative Deprivation and Tamil Separatism in Sri Lanka

The incongruities between the ecogeographical region and political boundaries have often led to conflict within a state or between states. When political and territorial boundaries are superimposed upon ecogeographical regions, the incongruities between these are noticed frequently. In some cases, several states occupy parts of a single-ecoregion, in other cases states encompass more than one ecoregion. In case of states with more than one ecoregion, conflict can be stimulated for two reasons. First, multi-ecogeographical states are more often multi-ethnic states and some ethnic groups within a state are often politically weaker than others, often lacking sovereignty over the territories they occupy. Second even if different ecoregions are ethnically homogenous they often support populations of different densities. A more populous region may have the power to control the resources of other regions to the extent of exploiting the resources of one ecogeographical region for the benefit of people who live elsewhere. A kind of inter-state colonialism takes place.[112] Over-exploitation and degradation of the region warrant calling it "a national sacrifice area".[113]

Sri Lanka provided an example of this kind of conflict in a small state. It has a wet zone centred in the Southwest, and a dry zone located in the east and north. The wet zone is inhabited mainly by Sinhalese, while most of the Tamil ethnic groups live in the dry zone. Sinhalese dominate the state government. Government sponsored water projects in the dry zone benefited mainly the Sinhalese, many of whom resettled in areas occupied by Tamils. Thus the dry zone being a national sacrifice area has been one of the complex roots of civil war in Sri Lanka.[114]

Two reasons can be accounted for this kind of conflict. One is the apparent incongruity between the ecogeographical region and the territorial boundaries within the state inhabited by different ethnic groups. This superimposition of ethnic demarcated territorial boundaries on the ecogeographical region has led to the exploitation of the resources and the environment of the region by the ethnic dominant Sinhalese group ruling the state at the expense of the other Tamil groups. Second, is the exceeding carrying capacity and the natural threshold of the region beyond its regeneration. In Sri Lanka, the dry zone has been overexploited beyond its capacity to benefit the Sinhalese at the cost of the Tamils.[115] There are a number of reasons and theories put forward for Tamil separatism in Sri Lanka .One significant cause being underplayed is the consequence of a grim situation aggravated endemically by the burgeoning population growth, and its pressures on scarce land. With the advent of the British, education came to be considered an avenue for survival in the view of the limits placed on land by the environment in terms of both productivity and increasing landlessness. Within Sri Lanka , the Tamils came to occupy extensive and dominant positions in the bureaucracy, universities and business, decreasing the population's competition over additional exploitable land.

Aside from other strategic steps, the Sinhalese majority's compelling poverty and unemployment pressurized their governments to take steps to thin down the Tamil dominance. Thus followed the imposition of Sinhala as the official language, constriction of university admission for Tamils and adoption of a new constitution devoid of any federal structure.[116] What came as harsh set back to the Tamils was the land colonization of Sinhalese in the Eastern province considered by Tamils as an integral part of their traditional homeland.[117]

From 1950 onwards, the Sinhalese government sponsored attempts at and encouraged settlement of Sinhalese from the south to the north central and north eastern dry zone. These canal colony settlers in larger irrigation schemes began to form a reason for discontentment among the Tamils.[118] This was considered a further attempt at cutting deep into the already limited and scarce Tamil resources. The main motive behind such a settlement was Sinhalese landlessness which was more acute than among the Tamils.[119] This Sinhalese colonization of the Tamil ancestral homeland not only brought about a drastic transformation of the demographic structure reducing Tamils to a minority in their own land, but also resulted in a dependent population coming under tension and rebellion due to scarcity of resources, made worse by the incoming settlement.

There is evidence to prove that the agrarian policies of successive governments and state sponsored colonization schemes have brought about a demographic transformation in the Trincomalee and Amparai districts. While the Sinhalese were mainly 3 per cent of the population in 1921, they rose to 33 per cent in 1981. Similarly in Amparai an increase of 78 per cent in Sinhalese population was seen by 1981.[120]

The Tamils were also affected by land colonization in other districts, viz. Mannar, Vavunia and Batticola. It is also alleged that the Sinhala fishermen are being inducted to threaten the very way of life and source of living of Tamil fishermen.[121]

Group Vs Group

The Tamil youth disenchanted with the political process which gave priority to Sinhalese interests resorted to violence, murder, robbery and assassination. Riots in 1983 provided the turning point. Started by the killing of 13 Sinhalese army men by Tamil militants, the Sinhalese in turn retaliated in Colombo and other parts of the island killing hundreds of Tamils.

Group Vs State

A group vs group ethnic conflict turned into a conflict between a group and a state which supported the Sinhalese majority. By early 1980s, Sri Lanka found itself in the midst of what can be termed a civil war or secession, as several militant Tamil organizations led by the youthful Liberation Tigers on Tamil Eelam (LTTE) resorted to violence as a means to remedy their plight. Nothing short of an autonomy or separate state was considered as conducive for safeguarding their interest. Guerrilla warfare on the part of Tamil militants and military crackdown on them by Sri Lankan became a regular phenomenon.

State vs State

A group vs. state conflict, a national problem of Sri Lanka was externalized with the influx of Tamil refugees in 1983 into India. Fleeing the military atrocities in Sri Lanka, the Tamil refugees crossed over into the ethnically similar state of Tamil Nadu in India. Tamil Nadu housing about 200,000 Sri Lankan Tamils has empathized with the cause of the Sri Lankan Tamils. Due to the rivalry between the DMK and AIADMK political parities in Tamil Nadu, each party has vied to be perceived as supporting fellow Tamils. Each is motivated both by genuine concern as well as by pure political pragmatism. The very dynamics of working of the Indian federal system have urged upon the centre not to minimize the Tamil cause, in the context of the state's penchant for independence. The ruling parities in Tamil Nadu and in New Delhi provided Tamil militants sanctuary, military and material help. Tamil Nadu was an important factor pressurizing New Delhi to support the Tamil insurgents. After the assassination of Prime Minister Indira Gandhi, Rajiv Gandhi tried to steer clear of interfering in the affairs of Punjab, Kashmir and Assam. In this attempt to settle the internal problems, the policy on Tamils of Sri Lanka changed. On failing to solve the issue peacefully the Indian government air lifted humanitarian aid in May 1987 with fighter jets as protection into the territory of Sri Lanka. The internationalization of the issue took place as Sri Lanka tried to seek the support of the United States and other powers.[122]

Subsequently, Rajiv Gandhi signed an agreement with President Jayawardene on 29 July 1987,which empowered India to militarily intervene to disarm militants and maintain a cease-fire with a quid pro quo for Sri Lanka promising a modicum of autonomy to Tamils in North East Sri Lanka. This outraged both the Tamil militants for not being satisfied with the deal and the Sinhalese for having conceded too much to the Tamils.

This aggravated the situation and widened the difference between the Indian government and Tamil militants. Rajiv Gandhi was assassinated in Tamil Nadu while electioneering by the Tamil militants. Premadasa, who succeeded Jayawardene, stressed that the Indian Peace Keeping Forces return to India. The ethnic conflict continues, despite assurances of solutions by successive Governments in Sri Lanka. At this stage the Sinhalese again started colonizing the eastern sector. With the deepening of the crisis, it has been ironic to note that just five years after the Indian troops were unceremoniously asked to quit Sri Lanka, the island's majority Sinhalese population, which cheered the withdrawl, wishes New Delhi would once again come to their country's rescue.[123]

As seen from the problems of environmental degradation and social effects in Sri Lanka, the resource base is gradually declining with land man ratio dwindling to an alarming extent. In a complex system of causes of ethnic conflict in Sri Lanka, environmental scarcity of resources appears to be one of its primary constituents. In the absence of a satisfying and enduring solution to the Tamil ethnic conflict, any military solution on the part of the Sri Lankan government is bound to arouse Tamil sympathy in India.In view of the stranglehold the Sinhalese Buddhists have had in Sri Lankan political decision making, the Sri Lanka government cannot give the Tamils more than what the Sinhalese want the government to concede.

As the age structure shows, large number of unemployed youths will enter the job market by AD 2010. The deteriorating economic base, declining environment and growing population would reinforce the Sinhalese masses' psychological and social sense of backwardness by the invidious comparison with the advanced Tamils in terms of disparaging evaluation of self worth which has often manifested in anxiety, violence and hostility.[124] In the absence of any truce satisfying the Tamil demands, hostility and violence that will increase between the Tamil militants and the Sri Lanka Government may compel many Tamils to migrate to Tamil Nadu in India posing security threats to India as the engagement of these Tamils in illegal activities like smuggling, and fratricidal murder of other militants have already become a law and order problem in the state. Tamil tigers appear to have a free run of the place in several villages on the Tamil Nadu coastline. As former Union Minister Chidambaram admitted, saying that the Tamil Nadu coastline is dotted with villages where the LTTE appears to have a free run.[125] The overwhelming task of settling Tamil refugees will be a strain on Tamil Nadu's economy and resources.

The Srilankan Tamil refugees came from a poor country to a poorer country like India. What were not easily available to the poor Indians were made available to the refugees at subsidized rates and also free of costs on humanitarian grounds. Approximately, 15 crores of rupees are spent annually on the maintenance of the refugees by Tamil Nadu government to be finally reimbursed by the Indian government.

Being prone to water scarcity,[126] refugee settlement will aggravate the crisis sometimes resulting in group fighting and conflict with the increasing military activities on the part of Sri Lanka government, the

Tamil Nadu government may be more empathetic to the Tamil cause than what central government would want Tamil Nadu to be. So India's evincing greater interests for these reasons in solving the crisis cannot be certainly "an independent Tamil Eelam which could have easily become a safe sanctuary and staging area for Sikhs, Assamese and other militant separatists in India.[127] The present continuing Sri Lankan military onslaughts on LTTE and establishing its stranglehold in North East peninsula has aroused strong reaction from the Tamil Nadu state of India with its government, people and opposition parities observing 'bandh' and asking the Indian government to intervene in order to stop decimation of innocent Tamil people by Sri Lankan military forces.

The Tamils in Sri Lanka lost Indian sympathy following the assassination of Rajiv Gandhi by LTTE. The Indian Government sincerely wants a political solution to the ethnic crisis in Sri Lanka. Continuation of the crisis has serious implication on Indian security, unity and integrity. From a recent interrogation of NSCN militants caught in Nagaland early November, the LTTE's well established connections within Bangladesh with insurgent groups in North East and the United Liberation Front of Assam (ULFA) in terms of military weapons supply and traning, were corroborated by the Indian Home Ministry to the greater security concerns of India.[128]

As found from the study of environmental degradation and various social effects, the people in Sri Lanka including the landless, rural and urban poor and the growing unemployed youth would be affected by an increasing sense of relative deprivation, caused by faulty government economic policies as well as environmental degradation and environmental scarcity of resources, it may result in situations manifesting violence, murder and conflict. Eventually. the government may depend on the Buddhists clergy to highlight the Tamil Sinhalese ethnic crisis, keeping in view the need,first,to reinforce their legitimacy in power and secondly to drive the depraved people's attention away from the immediacy of the situation by internationalizing the already existing ethnic crisis.

Environmental Degradation, Relative Deprivation, Ethnic Conflict in Pakistan

Ethnoregional sentiments have been growing stronger among the Pathans, Baluchis and Sindhis against the Punjabi dominated Pakistan.

This has increased ever since Pakistan attempted at Punjabization of all regions regardless of their distinctive culture and language with Punjabis ruling the roost in all spheres whether it is industry, military, bureaucracy or economic or politics.

Extensive influx of refugees from India (Muhajirs) during partition and from the other provinces of Pakistan to Sind was the primary reason to create feeling of deprivation and subordination in Sind people when those aliens contributed to the economic and commercial prosperity of Karachi without bestowing corresponding benefits on the indigenous Sindhis.[129] Their demands for more autonomy in terms of political and economic representation in military and bureaucracy remain still potential sources of conflict. The recurrence of riots in Karachi decimating Muhajirs may point to a sense of deprivation among the Sindhis. The migration to Karachi from other provinces has reduced the Sindhis to a minority.[130] Besides diversion of the Indus Water to Punjab at the cost of Sindh, the influx of people from India (Muhajirs) and from Punjab has resulted in abysmal decrease in land-man ratio, resource base and economic opportunities for the local people. To the resentment of Sindhis some of the best and most fertile land in the Ghulam Mahammad barrage area of Sind from which they have been ousted, have been allotted to the ex-servicemen who are mostly Punjabis. As pointed out, large tracts of vulnerable land in Sindh brought under irrigation since independence have been allotted to Punjabis, member of armed forces, senior bureaucrats and their relatives.[131] This, further shrinked the land for the local Sindhis. As a result of this infiltration, it is natural that it resulted in a conflict between the locals and the non-locals. Sindhis's initial demand for a homeland targeted at Muhajirs who contributed a lot to the economic and commercial expansion of Karachi. In the NWFP, the call for an independent entity of Pakhtunistan comprising the Pathan areas of both sides of the Afghanistan-Pakistan border, predates independence. Being extremely conscious of their distinct, ethnic and cultural identities, they are also strongly discontent with the growing dominance of Punjabi in all spheres to their despair and impoverishment. The politics of preference for Punjabi in matters of land and water as seen earlier, was the bone of contention between NWFP and Pakistan.

Baluchistan is Pakistan's largest and poorest land most sparsely settled province. It constitutes 40 per cent of Pakistan's landmass but only 5 per cent of its population. What is the bone of contention is that a majority of Baluchistan's population is non-Baluch. In the northern areas of the province, Pathans predominate and in Las Bela and Sibi districts, Sindhi and Siraiki are spoken more frequently than Balchi or Brohi, whereas in Quatta District, Punjabi or Pustho is more often spoken. The struggle of the Baluchis against what they call the tyranny of the Punjabi elite has been going on for the last three decades. The Sindhi elite's fear of discrimination in career advancement and other economic opportunities is similar to the threat which the locals of Baluchistan and NWFP feel from Punjabis, who have continually migrated to Quetta in Peshawar and to other cities.

A perusal of ethnic consciousness of the Pathanas, Baluchis and Sindhis will indicate a feeling of 'relative deprivation. This has existed all along among all the three groups. Without being sensitive to the development of these people, Pakistan policy of giving predominant position to Punjabis in education, service, bureaucracy, economic and military has created the relative deprivation among them. The ethnocentrism of Punjabis in Pakistan's political economy has given rise to sharpening the ethnic identities among these groups. What has been strongly opposed to are the assimilationist and redistributional policies. Given the dominance of Punjabi, the assimilationist policies are viewed as clever subterfuge to weaken the minority and strengthen the Punjabi domination. Its distributional policies are resented because by these, the majority community redistributes resources to minority communities.[132]

Besides, with Bangladesh being independent, "Pakistan has been the victim of a successful secessionist movement and 'would be opponents' of the continued integrity of Pakistan are encouraged by their predecessor's success".[133]

To these feelings of 'relative deprivation' among the disgruntled in Pakistan will be added equally the deteriorating environmental and resource depletion in these regions. Growing population, its pressures on land which has already been degraded due to various reasons like siltation, salinity and water scarcity, urbanization, transmigration of people from rural to urban, competing with them for the basic human needs, need for more productivity on marginal land and dependence on fertilizers, for this, are some of the grave problems afflicting these regions. 27 million hectares of land in Pakistan are already affected by desertification. Water scarcity, desertification and degradation of land have had deleterious effects on productivity on which teeming millions depend. The Kalbagh dam project which will displace 250,000 and submerge vast areas in NWFP and Baluchistan, with adverse effects on the Indus delta system in terms of increasing salinity, siltation and water scarcity has been a subject of conflict between the central government and the regions. Though the Indus Water Apportionment Accord has been signed distributing water among these regions, a further considerable reduction in the flow of Ganges water will happen due to this project and will aggravate the resentment of these regions against the Punjabi dominated Pakistan. The influx of 3.4 million Afghan refugees as a result of the Afghan war into NWFP of same ethnic background, i.e., Pushtun, with their 3 million livestock, apart from having ethnic rivalry, law and order problems and a drain on scarce resources, have degraded the environment in terms of overgrazing, deforestation and erosion of water courses. It is alleged that due to water courses being eroded, the lifespan of the important Mangla dam is estimated to be halved with other concomitant-disastrous consequences for irrigation and power generation.[134]

Link Between Urban growth and Violence

Between 1950 and 1990,there was a fivefold increase – to 1.5 billion– in the number of urban residents in developing countries; about 37 per cent of the population of the developing world now lives in cities. By 2025, the United Nations projects a future tripling of the total to 4.4 billion, at which point nearly two thirds of the citizens of the developing world will live in cities.[135] Apart from natural growth. reclassification, contribution of permanent migration from rural areas to urban growth is huge, though below 50 per cent.[136] Rural – urban migrants tend to be relatively young which accentuates the youth bulge in urban population in poor countries. This age distribution contributes to natural population growth in cities, because a younger population has more children and fewer deaths compared to an older population.[137]

During the 1960s, some scholars suggested that rural to urban migration in the developing world could cause a spurt in urban violence. According to Wayne Cornelius there are three hypotheses on this linkage.[138] First, migration could breed economic frustration, migrants' expectations could not be fulfilled and the exorbitant and inordinate consumption of the elites, could enhance the migrants' awareness of their marginal status. Second, cultural incongruity, conflicts and disruption of past living habits and custom of the migrants would cause personal identity crisis. This would increase the chance of primary group breakdown and weakening traditional social controls on deviant behaviour. As migrants sought entry into new protective groups, they would be easily recruited into extremist political activity. Third, the urban environment provides an opportunity for high levels of social communication, increased political awareness and mobilization of radical opposition to go hand in hand. This would politicize migrants and encourage mass involvement in political action.[139]

Little empirical research has been done in recent years on this to prove these hypotheses. It needs to be studied whether the links are strengthened when large migrants interact with many factors like economic crisis, poverty and environmental scarcity of resources.

Pakistan's Cities Such as Karachi- An example of this Linkage

In Pakistan, the migration from rural areas to urban areas has been constantly rising. A growing number of shanties are seen around Karachi. As predicted, sea level rise submerging Karachi may result in flight of people for food and shelter to relatively secure urban areas and other places. The new level of conflict will spawn up among various groups. Competition for jobs, economic benefit and basic modern amenities among the urban population will entwine migrants and the original inhabitants in various conflicts. Government's attention is concentrated on urban areas and for meeting their basic modern demands such as provisions for water, education, electricity, health and sanitation by diverting a sizeable part of the total budget. The disillusioned rural people and poor urban people will be thus disenchanted with the government setting the stage for political instability. An analyst of rural urban relationships in developing countries describes the conflicts that may arise thus.

That most important class conflict in the poor countries of the world today is not between labour and (those who control) capital, nor is it between foreign and national interests. It is between the rural classes and urban classes. The rural sectors contain most of the poverty and most of the low cost sector of potential advance. But the urban sector contains most of the articulateness and power... As a result, the urban classes have been able to win most of the rounds of struggle with the country side; but in doing so they have made the development process needlessly slow and unfair.[140]

Karachi grows by about 400,000 people per year. Where as its population rises at around 5 per cent per annum, urban services expand by only 1.2 per cent.[141] Government and development authorities cannot provide residents with basic services; the city's aging infrastructure is over taxed and does not properly service new communities. Karachi has been plagued by housing problem, acute shortages of electricity, and water, disarranged transport system and trafficking in narcotics and arms. High demand for water along with rampant corruption and mismanagement in the Karachi Water and Sewage Board has created a tanker Mafia cornering all the profits by selling water to customers at exorbitant prices, often on credit.[142] Frustration stemming from the lack of urban services has prompted attacks on the offices of the Karachi Electricity Supply Corporation and the Karachi Water and Sewage Board.[143] The inability of state institutions to address diverse demands accentuate latent ethnic and class tensions. Immigration and high urban population growth further boost grievances, highlight the impotence of the state, and reduce state legitimacy. This strong urban bias in terms of provision of services such as education, health, electricity and water widens social inequities and increases social conflicts. Though some studies show little historic correlation between rapid urbanization and civil strife,[144] Pakistan has recently been affected by an orgy of riots in Karachi.

As evident from the above analysis, the environmental degradation may not be root cause of ethnic disputes in Pakistan but however, appear to exacerbate the problems. The existing feelings of relative deprivation among various ethnic regions in Pakistan may be further aggravated by the environmental degradation along with its social effects. In the case of such ethnic crisis and rise of separatism in Pakistan, the Pakistani government being unable to address these problems will find a convenient hypothesis to raise the bogey of Indian involvement as in recent riots in Karachi, Pakistan alleged Indian role.[145] The Indian involvement was emphasized to secure legitimacy by the ruling elite in its domestic front.

This would gain support of its fundamentalist Muslim population. These as usual orchestrated allegation of Indian premeditated plan to dismember Pakistan are believed to have happened in case of Bangladesh by the Pakistani government. This factor could potentially bedevil Indo-Pak bilateral relations.

The unchecked build up of carbon dioxide and other greenhouse gases in the global atmosphere due to industrialization would cause, as predicted, far reaching disruption in temperature and rainfall patterns.[146] In the previous chapter on social effects of environmental degradation the impact of global warming on agricultural production and the overall economy of developing countries has been studied. As predicted, in agricultural terms Pakistan would find itself worse off than India – in turn possibly affecting their relations.[147] The prospects of food insecurities and declining economy for a growing population would very much question the credibility and legitimacy of Pakistani state and government to rule. Further the existing ethnic crisis and separatist tendencies among various regions will be exacerbated by this growing sense of relative deprivation among the poor, deprived and depraved people. Pakistan in order to divert domestic attention could draw India into the conflict.

The Philippines has also demonstrated clear links between ecological marginilazation and conflict. According to Daniel Lascon, in the early 1990s there are two sources of poverty and injustice behind the country's chronic insurgency: the accumulation of land in the hands of a few who failed to deal with the problems of poor, and land degradation that affected small farmers and was not alleviated by government action.

References

1. John Dollard et al. Frustration And Aggression (New Haven, Conn.: Yale University Press, 1939); Leopard Berkowitz, Aggression: A Social Psychological Analysis (New York: McGraw-Hill, 1962). On relative deprivation, see James Davies, "Toward a Theory of Revolution ", American Sociological Review 27, no.1, 1962, pp. 5-19; Ted Gurr, Why Men Rebel (Princeton, N.J.: Princeton University Press, 1970); and Ted Gurr and Raymond Duvall,"Civil Conflict in the 1960s: A Reciprocal Theoretical System with Parameter Estimates", Comparative Political Studies Vol. 6, no.2, 1973, pp. 135-69.

2. Henri Tajfel, ed., Differentiation between Social Groups (London: Academic Press, 1978); Edward Azar and John Burton, International Conflict Resolution: Theory and Practice (Sussex: Wheatsheaf, 1986); Lewis Coser, The Functions of Social Conflict (London: Free Press, 1956) and Donald Horowitz, Ethnic Groups in Conflict (Berkeley: University of California Press, 1985).

3. See Thomas F. Homer Dixon. Environment. Scarcity and Violence (Princeton/ New Jersey: Princeton University Press. 1997). p. 137.

4. Alexander Wendt. "The Agent-Structure Problem". International Organisation Vol.41, no.3, 1987, pp. 335-70, and Anthony Giddens. The Constitution of Society: Outline of the Theory of Structuration (Berkeley: University of California Press, 1984).

5. See Kenneth Waltz, Theory of International Politics (Reading, Mass: Addison-Wesley, 1979); and Bruce Bueno de Mesquita, The War Trap (New Haven, Conn.: Yale University Press, 1981).

6. Thomas Homer Dixon, Environment, Scarcity and Violence, p. 137.

7. Charles Tilly, From Mobilization to Revolution (Reading, Mass: Addison-Wesley,1987); and Sidney Tarrow, Power in Movement : Social Movements, Collective Action and Politics (Cambridge : Cambridge University Press, 1994).

8. Volker Boge, "Proposal for an analytical framework to grasp environmental conflict", Environmental Conflicts Project(ENCOP), Centre for Security Studies and Conflict Research, Swiss Federal Institute of Technology, Swiss Peace Foundation, Zurich./Bern, Occasional paper no. 1, July 1992, p. 4.

9. Nazli Choucri and Robert North, Nations in Conflict (San Francisco : Freeman, 1975).

10. For a general analysis of the role of scarcity of nonrenewable resources in interstate conflict, see Ronnie D. Lipschutz, When Nations Clash (Cambridge, Mass.: Ballinger Publishing, 1989).

11. Waltz, Theory of International Politics; Robert Gilpin, War and Change in World Politics(Cambridge : Cambridge University Press,1981): For views of political change in realist theory, see R.B.J Walker, "Realism,Change and International Political Theory", International Studies Quarterly Vol. 31, no. 1, 1987, pp. 65-86

12. See Robert Albion, Forests and Sea Power: The Timber Problem of the Royal Navy,1652-1863(Cambridge,Mass ; Harvard University Press ,1926).

13. Thomas F. Homer Dixon, Environment,Scarcity and Violence, p. 139.

14. Peter Gleick, Water and Conflict, Occasional paper no.1, Project on Environmental Change and Acute Conflict (Cambridge, Mass: American Academy of Arts and Science and the University of Toronto, 1992); Peter H.Gleick, World's Water: The Biennial Report on Fresh Water Resources (Washington D.C.: California, Island Press, 1998), pp. 105-153.

15. Ismail Serageldin, "Earth Faces Water Crisis", Press Release, World Bank, Washington, D.C.: 6 August 1995.

16. Quoted in Mikiyasu Nakayama, "Water : The 21st Century Oil," Work In Progress, A Review of Research Activities of the United Nations University Vol. 15, No. 2, Winter 1998, p. 18.

17. Peter H.Gleick,World's Water, p. 107

18. United Nations, Comprehensive Assessment of the Freshwater Resources of the World.Prepared for the Commission for Sustainable Development and the General Assembly Stockholm Environment Institute and the United Nations, New York 1997; P.H. Gleick, "Water and Conflict", International Security Vol. 18, no. 1, pp. 79-112.

19. M.R.Lowi, Water and Power : The Politics of a Scarce Resource in the Jordan River Basin (Cambridge : Cambridge University Press, 1995) pp. 115-144.

20. *Ibid.*, p.125

21. Miriam R.Lowi, "Water and Conflict in the Middle East and South Asia : Are Environmental Issues and Security Issues Linked?, The Journal of Environment and Development", Vol.8, No. 4, December 1999, pp. 387-88.

22. S.Chira, "North Korea dam worries the South", The New York Times, 30 November 1986, p. 3 ; and N. Koch, "North Korean dam seen as potential water bomb", Washington Post/ San Francisco Chronicle , September 30, 1987, p. 3

23. P.H. Gleick, "Water and Conflict", *op.cit*, pp. 79-112.

24. A.Dupont, "The Environment and Security in East Asia", The Adelphi Paper, Research School of Pacific and Asian Studies, Strategic and Defence Studies Centre, Australian National University, Canberra, Australia, 1998.

25. J.Lee, "KL uses water as a political tool", The Jakarta Post, 7 July 1997.

26. G.P.Zachary, "Water Pressure : Nations Scramble to defuse fights over supplies", The Wall Street Journal, 4 December 1997, p. A17.

27. Peter H. Gleick, World's Water, p. 111.

28. Marq De Villiers, Water Wars (London ; Weidenfeld and Nicolson,1999), pp. 239-40.

29. Quoted in Ibid, p. 247; See also Norman Myers, "Environment and Security" Foreign Affairs, no.74, Spring 1989, p. 32; and Thomas Naff and Ruth Matson, eds., "The Nile River", Chapt.6 in Water in the Middle East : Conflict or Cooperation? (Boulder , Colo . : Westview Press, 1984), pp. 125-55.

30. Quoted in Marq De Villiers, p. 251.

31. John Kolars and William Mitchell, The Euphrates River and the South East Anatolia Development Project (Carbondale : Southern Illionis University Press,1991).

32. Homer Dixon, Environment, Scarcity and Violence, p. 140.

33. Quoted in Marq De Villiers, p. 254.

34. Francoise Chipaux, "Syria and Iraq unite against Turkish Dam," Manchester Guardian Weekly, 17 February 1996.

35. German Advisory Council on Global Change Annual Report 1997, World in Transition : Ways Towards Sustainable Management of Fresh Water Resources, Springer : Varlage Berlin Heidelberg, 1999, pp. 208-209.

36. Marq De Villiers, p. 261.

37. Marq De Villiers, p. 227

38. World in Transition, p. 210.

39. World in Transition, *op.cit.*, p. 210.

40. T. Stauffer, "The Price of Peace, the Spoils of War", American - Arab Affairs,1982 Summer, p. 1.

41. Miriam R. Lowi, Water and Conflict in the Middle East and South Asia, p. 388.

42. Marq De Villiers, pp. 229-30.

43. Ibid, p. 230.

44. Miriam R.Lowi, Water and Conflict in the Middle East and South Asia, p. 388.

45. M.R.Lowi, "Rivers of Conflict, Rivers of Peace", Journal of International Affairs Vol. 49, no. 1, 1995, pp. 125-127.

46. M.R.Lowi, Water and Conflict in the Middle East and South Asia, p. 385.

47. P.Heyns, "Existing and Planned water development Projects on international rivers within the SADC regions", presented at the conference of SADC Ministers Water Resources Management in Southern Africa : A Vision for the Future, Pretoria, South Africa, 23 November 1995.

48. J.Leitch, "Okavango, a disputed river", Hydroplus : International Water Review Vol. 77, 1997, pp. 20-24.

49. *Ibid.*

50. J.Arenstein, "A river barely runs through it", Mail and Guardian African Eye News, South Africa, October 13, 1996.

51. D.Lilienthal, "Another Korea in the making?", Collier's Magazine, April 4, 1951, p. 58.

52. Ian Stephens, Pakistan, Old Country-New Nation (Australia : Penguin Books, 1964), p. 276.

53. Earthscan, Environment and Conflict, Earthscan Briefing Document, 40 London : International Institute for Environment and Development, 1984, p. 55.

54. P. Gizewski and Thomas F.Homer Dixon, Environmental Scarcity and Violent Conflict: The Case of Pakistan. In the Project on Environment, Population and Security, Peace and Conflict Studies, Toronto, Canada: University of Toronto, 1996.

55. Hindustan Times (New Delhi), 6 November 1998.

56. David G.Lemarquand, International Rivers : The Politics of Cooperation (University of British Columbia and the Waterloo Research Center, 1977).

57. India, Preservation of Port of Calcutta Report . (New Delhi : Government of India Press.1961) February 22, 1961, p. 5.

58. See Nahid Islam, "The Ganges water Dispute : Environmental and Related Impacts on Bangladesh", BIISS Journal Vol. 2, no. 3, July 1991, pp. 263-92.

59. See for details of Conflict, negotiations and talks between the two countries, N.Gann, Environment and National Security: The Case of South Asia (New Delhi: South Asian Publishers, 2000) pp. 141-156; Ben Crow, Alan Lindquist and David Wilson, Sharing the Ganges: The Politics and Technology of

River Development (New Delhi: Sage publication, 1995); Khurshida Begum, Tension over the Farakka Barrage: A Techno-Political Tangle in South Asia (Dhaka: University Press, 1987).

60. See for the opinion of V.Surynarayan, Director, Centre for South and SouthEast Asian Studies, Madras University and T.S. Rama Rao, Professor of International Law, Madras University, T.S. Subramaniam, "Island of Contention : Kachchativu and Fishermen's rights", Frontline Vol.11,No.8,9-22 April 1994, pp. 49-50.

61. See S.D.Muni, "Kachchathivu Settlement : Befriending Neighbouring Regimes", Economic and Political Weekly Vol. 9, no. 28, 13 July 1974, pp. 119-22; See also Ashok Varma, The Government of India's Approach to Territorial Boundary Disputes : The Case of Kachchativu (Thesis : B.Phil, St. John's College, University of Oxford, 1976).

62. Foreign Affairs Record (New Delhi : Ministry of External Affairs, 1974), Vol. 20, p. 227.

63. *Ibid.*

64. Quoted in Frontline, *op.cit.*, p.50. see also for Swaran Singh's statement in the Lok Sabha on 23 July 1974, Foreign Affairs Record, *op.cit.*, p. 226.

65. See Y.B.Chavan's statement in Indo-Sri Lanka Maritime Boundary Agreement, Foreign Affairs Record (New Delhi, Ministry of External Affairs, 1976) Vol. 22, p. 91.

66. See for this kind of Indian thinking, Frontline, *op.cit.*, pp. 50-51.

67. Frontline, *op.cit.*, p. 51.

68. Sunday Magazine (Calcutta), 22 September 1991.

69. Ibid.

70. Don Hinrichsen, Our Common Seas : Coasts in Crisis (London : Earthscan Publication Ltd in association with United Nations Environment Programme, Nairobi,1990), Chapter 9, South Asia, pp. 121-131.

71. Frontline, *op.cit.*, p.49.

72. The Sunday leader, quoted in Down to Earth Vol. 14, No. 11, 31 October, 1995, p. 8

73. See N.Gaan, "Environmental Scarcity of fish and Conflict : The case of India and Sri Lanka over Kachchativu Waters", International Studies Notes Vol. 22, no. 3, Fall 1997, pp. 17-23.

74. Sanjoy Hazarika, Bangladesh and Assam : Land Pressures, Migration and Ethnic Conflict, Occasional Paper Series of the Project on Environmental Change and Acute Conflict, A Joint Project of the University of Toronto and the American Academy of Arts and Sciences, No.3 March 1993, p.46.

75. Towards Sustainable Development : National Conservation Strategy of Bangladesh , Ministry of Environment and Forest, Government of Bangladesh/IUCN, 1991, p. 69.

76. Mohabab Hossein, "Briefing Paper (Bangladesh)" (London : Overseas Development Institute, November 1990), p. 1.

77. *Ibid.*, p. 3.

78. Atiur Rahman, Impact of Riverbank Erosion : Survival Strategies of Displacees (Dhaka : Bangladesh Institute of Development Studies, 1985), p. 11.

79. John R.Rogge, Riverbank Erosion : Flood and Population Displacement in Bangladesh (Dhaka : Jahangirnagar University ,October 1990). p. 35.

80. Atiur Rahman, *op.cit.*, pp. 19-20.

81. Paul Harrison, The Third Revolution : Environment, Population and a Sustainable World (London/ New York : Tauries and Co. Ltd., 1992), p. 230.

82. Don Hinrichsen, Our Common Seas : Coasts in Crisis (London : Earthscan Publication Ltd. In association with United Nations Environment Programme, Nairobi, 1990), chapter 9, p. 127.

83. Jessica Tuchman Mathews, "Redefining Security", Foreign Affairs Vol. 68, no. 2, Spring 1989, pp. 162-177.

84. Thomas F.Homer Dixon, "Environmental Change and Economic Decline in Developing Countries", International Studies Notes Vol. 16, no. 1, Winter 1991, p. 22.

85. Sanjoy Hazarika, *op.cit.*, p.52.

86. *Ibid.*

87. Shikha Mukharjee, Special Correspondent of Times of India, Calcutta, December 1992, quoted in Sumit Ganguly, "Ethno-religious Conflict in South Asia", Survival Vol. 35, no. 2, Summer 1993, p. 100.

88. Raj Chengappa and Ramesh Menon, "The New Battle Fields", India Today, 31 January 1993, pp. 28-29.

89. Sumit Ganguly, "Ethno-religious Conflict in South Asia", Survival Vol.35, no.2, Summer 1993, p.100.

90. Times of India, 26 November 2001; Samaj, November 1 2001 ; Pragatibadi, 28 September 2001.

91. Prajatantra, 14 January 2001 ; Sambad, 2 December 2000 ; Sambad, 13 February 2001.

92. Robert G.Gosselink, "Minority Rights and Ethnic Conflict in Assam, India", Boston College Third World Journal Vol.XIV, Winter 1994, p.91.

93. *Ibid.*

94. Sharifa Begum, Population, Birth, Death and Growth Rates in Bangladesh : Census Estimates (Dhaka,Bangladesh Institute of Development Studies), 1990, p. 56; Marcus Franda,Bangladesh : The First Decade (New Delhi : South Asian Publishers,1981) p.235 ; Census of India, 1991.

95. Amiya Kumar Das, Assam's Agony : A Socio-Economic and Political Analysis. 1982, p.38, quoted in R.G. Gosselink, *op.cit.*, p. 90.

96. Sanjoy Hazarika, Assam : A Crisis of Identity (1980), p. 10, quoted in Gosselink, Ibid. : see also Myron Weiner, Sons of the Soil : Migration and Ethnic Conflict in India (New Delhi : Oxford University Press, 1988) pp. 80-81.

97. Thomas F. Homer Dixon, "Environmental Scarcities and Violent Conflict : Evidence from Cases," International Security Vol. 19, no. 1, summer 1994, p. 22; see also Sanjoy Hazarika, *op.cit.*, pp. 52-54.

98. Prasun Sonwalkar, "Exodus of Bangladeshis to India", Times of India , 6 December 1991, p. 4; Amulya Ganguli ," Bangladesh Lebensraum", Times of India, 13 July 1992.

99. Sanjoy Hazarika, "India Admits Failure to Cut Bangladesh Influx", New York Times, 16 December 1992,

100. Sanjoy Hazarika, *op.cit.*, p.58.

101. *Ibid.*, p.59.

102. Ruben Banarjee, "The NorthEast : Returns to Arms", India Today, 30 November 1992, pp.36-37.

103. Jonathan S.Landay, "Gandhi Loses a Election to Party Favouring Expelling Moslems", UPI, 18 December 1985 available in Lexis, Nexis Library, UPI File.

104. Mani Deb, "Assam's Patience with Delhi Grows Increasingly Thin", Financial Times, 3 February 1988.

105. Times of India , 29 August 1995, p. 6

106. Times of India, 1 December, 1995. See also Nitin A. Gokhale, "Power Keg : What makes the North-east of the Country's Hottest Trouble Spots for Alleged Hands of External Powers like Bangladesh and Pakistan", Sunday Vol. 23, no. 12, 24-30 March 1996, pp.14-22.

107. Sanjoy Hazarika, "Bangladesh and Assam : Land Pressures", *op.cit.*, p. 61.

108. Homer Dixon, Environment, Scarcity and Violence, p. 142.

109. Kurt Schock, "A Conjunctural Model of Political Conflict", Journal of Conflict Research Vol. 40, no. 1, 1996, pp. 93-133.

110. Mark Lichbach, "An Evaluation of Does Economic Inequality Breed Political Conflict Studies", World Politics Vol. 41, no. 4, 1989, pp. 431-70; See also Raj Desai and Harry Eckstein, "Insurgency : The Transformation of Peasant Rebellion", World Politics Vol. 42, no. 4, 1990, pp. 441-465.

111. Peter Wallensteen, "Food Crops As a Factor in Strategic Policy and Action", in Arthur H. Westing ed., Global Resources and International Conflict : Environmental Factors in Strategic Policy and Action (New York : Oxford University Press, 1986), pp.154-55.

112. Bruce Byers, "Ecoregions, State Sovereignty and Conflict", Bulletin of Peace Propasals Vol. 22 (1), 1991, p.69.

113. See Christoper McLeod, Randy Hayes and Glenn Switkes, " The Four Corners : A National Sacrifice Area", reprinted from The Workbook Quarterly Journal of the South West Research and Information Centre. P.O. Box 4524, Albuquerque NM.87106,USA . April-June 1985.

114. See Chelvaduari Manogaran, Ethnic Conflict and Reconciliation in Sri Lanka (Honolulu: University of Hawaii Press, 1987).

115. See Bruce Byers, *op.cit.*, p.71.

116. Robert N.Kearney , " Ethnic Conflict and Tamil Separatist Movement in Sri Lanka", Asian Survey Vol. 25, no. 9, September 1985, pp. 904-905; James Manor, ed., Sri Lanka in Change and Crisis (New York : St Martin Press, 1984), p. 1

117. Mohan Ram, Sri Lanka : The Fractured Nation (New Delhi : Penguin Books, 1989), p. 33.

118. K. M. de Silva, Managing Ethnic Tensions in Multi-Ethnic Societies : Sri Lanka, 1880-1985(Lanham, MD : University Press of America ,1986); C. Manogaran, Ethnic Conflict and Resolution in Sri Lanka (Honolulu, HI : University of Hawaii Press,1987).

119. S.V. Kodikara, "The Separatist Eelam Movement in Sri Lanka : An Overview", India Quarterly Vol. 37, April-June 1981, pp. 194-212.

120. M.Y.Mohammad Siddique, "Land and Colonization Problems in Aparai Districts", Cintanai (Jaffana) Vol. 1, July 1983, pp. 94-212.

121. V.Suryanarayana, "Ethnic Conflict in Sri Lanka", in Urmila Phadnis, S.D. Muni and Kalim Bahadur, ed., Domestic Conflicts in South Asia, Vol. 1, Political Dimensions (New Delhi: South Asia Publishers, 1986), p. 140.

122. See, New York Times, June 1987, p. 9.

123. Times of India, 11 May 1995.

124. Donald Horowitz, Ethnic Groups in Conflict (Berkeley and Los Angeles : University of California Press, 1985), p.167.

125. Quoted in Times of India, 4 September 1995.

126. Paul P.Appasamy, "City of Thrist", Down to Earth Vol.2, No.20, 15 March 1994, pp. 34-35.

127. Marshall R.Singer, "Prospects for Conflict Management in the Sri Lankan Ethnic Crisis", in Joseph V. Montville, ed., Conflict and Peace Making in Multi-Ethnic Societies (New York : Lexington Books, 1991), p. 277.

128. "LTTE has Major Role in Arms Supply to Militants: Government", Times of India, 6 December 1995, p. 9.

129. Charles H.Kennedy, "The Politics of Ethnicity of Sindh", Asian Survey Vol. 31, No. 10, October 1991, p. 9.

130. See Shahid Jaded Burki, "Migration, Urbanization and Politics in Pakistan" in Howard Wriggins and James Guyot, ed., Population and Politics of South Asia (New York, 1983), pp.147-148, quoted in Uma Singh, "Ethnic Conflicts in Pakistan : Sind as a Factor in Pakistani Politics" in Urmilla Phadnis, S.D. Muni and Kalim Bahadur, ed., Domestic Conflicts in South Asia, op.cit., p.156.

131. Nahid Islam, "Ethnicity and Environment in South Asia", p. 53.

132. Donald Rothchild, "State and Ethnicity in Africa : A Policy Perspective" in Nevittee and Kennedy, Ethnic Preference and Public Policy in Developing States (Boulder,Co: Lynne Rienner Press, 1986). pp. 47-48.

133. Craig Baxter, Yogendra K.Mallik, Charles H.Kennedy and Robert C.Oberst, Government and Politics in South Asia (Boulder and London : Westview Press, 1987), p. 185.

134. Sandy Gordon, "Resources and Instability in South Asia", Survival Vol. 35, no. 2, Summer 1992, p. 76.

135. John Kasarda and Allan Parnell, "Introduction: Third World Development Issues" in John Kasarda and Allan Parnell ed., Third World Cities: Problems, Policies and Prospects (New York : Sage 1993), pp. IX-X.

136. A.S.Oberai, Population Growth, Employment and Poverty in Third World Mega-cities (New York : St. Martin's Press, 1993), p. 26.

137. *Ibid.*, p. 28.

138. Wayne A.Cornelius, Jr., "Urbanisation as an Agent in Latin American Political Instability: The Case of Mexico", American Political Science Review 63, no. 3, 1969, pp. 833-57.

139. Kenneth F.Johnson, "Causal Factor in Latin American Political Instability" Western Political Quarterly 17, no. 3, 1964, p. 441.

140. Michael Lipton, quoted in Lestern R.Brown and Jodi L.Jacobson, "The Future of Urbanization : Facing the Ecological and Economic Constraints", World Watch paper 77, May 1978, p. 13.

141. Farida Shaheed, "The Pathan-Muhajir Conflict, 1985-6: A National Perspective", in Veena Das, ed, Mirrors of Violence: Communities, Riots and Survivors in South Asia (New York: Oxford University Press, 1990), p. 205.

142. Ahmed Rashid, "Mean Streets : Chaos and Violence Rule in Karachi", Far Eastern Economic Review 157, no. 43, 27 October 1994, p. 19.

143. Arif Hasan, "Karachi and the Global Nature of Urban Violence", Urban Age 1, no. 4, 1993, p. 5.

144. Wayne Cornelius Jr. "Urbanization As an Agent in Latin American Political Instability : The Case of Mexico", *op.cit.*; Abdur Lodhi and Charles Tilly, "Urbanization,Crime and Collective Violence in 19th Century France", American Journal of Sociology Vol. 79, no. 2, September 1973, pp. 296-318.

145. Times of India, 2 January 1995, "Pakistan Cries Wolf Again", M.B. Naqvi from Karachi, p. 15.

146. See W. W. Kellog, "Society, Science and Climate Change", Foreign Affairs Vol. 60, 1982, pp. 1076-1109.

147. Norman Myers, "The Environmental Dimension to Security Issues", *op.cit.*, p. 253.

7

Environmental Security not an Appendage to Geo-Poltical World Order of the United States : A Critique

The cold war understanding of security was largely premised on the threats that states pose to each other in a system supposed to be fundamentally anarchic in the sense that political life therein is pontificated by states not subject to any authority superior to their own sovereignty. The concept of security is closely tied with the enhanced military capability of states making them impregnable against external or outside threats to maintain their independent identities, as well as their physical and functional integrity. Since there is lack of an international sovereign authority to circumscribe the ambitious and aggressive designs of other states, the states used to depend on themselves building up their own military capabilities to achieve security. These self security concerns of states often result in a destabilizing arms race, hostile or potentially hostile interstate relations or organized, intentional state directed violence through military means. In an unequal world these self help security concerns could lead to the emergence of an international system characterized by super power rivalry, and dominance, bipolarity, balance of power and arms race among the states.

With the demise of the cold war and disappearance of Soviet Union there has been much debate about how to reformulate concepts of national and international security, since the super power rivalry no longer occupies the center stage of policy decisions and global politics. What was overlooked and unheeded was the potential future threats of politico-socio-economic disruption springing from the environmental degradation. Advocacy of environmental degradation as non-military threats by Norman Myers, Jessica Tuchman Mathews, Richard Ullman and Michael Renner added grist to the already going on debate on rethinking security in the current changing international scenario.[1] Giving primacy to the environmental dimension in the security agenda in a more comprehensive fashion is what is termed as environmental security encompassing all in a global scale. This has questioned the very traditional military thinking, and put forth a strong argument for linking certain environmental problems with the prospects for political frictions and tensions or even war and peace. At the centre of the ongoing debate is the growing assertion that environmental scarcity or degradation is an important factor contributing to political instability or violent conflicts at local, regional and international levels. This kind of rethinking of security in terms of environment has emanated from a growing admission to the fact that a plethora of new and large scale environmental threats with political ramifications are beginning to dominate international discourse, debates and policy decision about environmental concerns.

The thesis that environmental degradation is a far greater non-military threat to entire human kind with consequences more devastating than the military threats even the nuclear has assumed a considerable rhetorical force. All these arguments emphasize the need to give primacy to environmental issues and include them at the top of the security agenda of national and international politics. As adduced by the scholars, the ramification, perils and social effects of environmental scarcities and degradation such as land degradation, soil erosion, deforestation, desertification, ozone layer depletion, industrial pollution and climate change in terms of political and economic difficulties such as economic decline, poverty, displacement of people, disruption in political institutions, are suggestive of the imperative to think in terms of environmental security and its world wide applicability to the global environmental problems which seem to be threatening all state and people.[2] Unlike the military threats requiring a bayonet to face a bayonet, the environmental security needs on the other hand international cooperation and efforts to tackle these impending environmental problems.

Environmental Security : Meaning and Nature

The environmental security is more comprehensive and all encompassing. The cold war understanding of security was confined to the geopolitical, economic and military interests of the dominant powers. What environmental security suggests a transcendence of the narrow focus on geopolitical interests of the dominant players. It is not just an addition to national and international security like political, economic or military. It envisages a transfiguration of the very substructure of what constituted the political and economic order of the superpowers created during the cold war. The above arguments rely on the environmental causes as the key to rethinking the political and economic metaphysics of the superpowers. "The term environmental security has become part of a framework for advocating new conceptions of how the international political order should be understood and what normative aspirations are appropriate in the post clod war period".[3]

The 1994 United Nations Development Program Human Development Report listed environmental security as one of the seven components of its formulation of global human security. In 1995 the Commission on Global Governance followed this line of argument to add planetary security to its reformulation of the post-Cold war security agenda[4]. In 1997 the Clinton administration adopted environmental themes as a focus of its foreign policy. The Woodrow Wilson Centre had undertaken a major project to examine the environmental dimensions of foreign policy.[5] These suggest clearly a political exercise about whose issues are part of the international policy agenda. These intended just to add environmental dimensions to their respective foreign policy agenda. It is said environmental security or ecological security is imprecise, and ambiguous. It is too much based on broad generalizations without having established sometimes the clear-cut links between environment and specific entities that may be rendered insecure. Thomes Homer Dixon, a wide known authority on environmental degradation and conflict points out how environmental scarcity of resources or their degradation has led to violent conflict within states or among states in some specific cases. The emphasis Homer Dixon tries to put is on causes and consequences of environmental scarcity of resources that have led to such violent conflict but surely not on the basis and existing pattern of economic and political thinking - the sole cause of the environmental degradation and consequently the conflict.

The debate on environmental security while continues emphasizing the need to redefine the conventional security, " is in danger of overlooking important political issues unless analysts are alert to the persistent dangers of the traditional ethnocentric and geopolitical assumptions in Anglo-American security thinking".[6] Though analysts have wakened the policy makers to the impending dangers of the linkage between environmental degradation and conflict citing many cases in South countries, and putting forth a strong case for rethinking security after the end of cold war, but no where is seen among the policy makers the efforts to rethink the ethnocentrism of western security paradigms. Environmental security is just tagged on to the traditional security thinking. A high pitched argument for environmental security continues without the traditional Anglo-American security thinking being asked for an overhaul. The debate remains part of academic discussions as well of international politics and foreign policy of nations. Environmental security has been emphasized upon to extend the foreign policy agenda of nations, just by another item, a part of their foreign policy and, not their security agenda, a part of the environmental security. Environmental security has been held as synonymous with North's political, economic or resource security, nothing more.

During the cold war the US and its allies' interests were carried as the interests of the entire globe. Any kind of threat to their interests was considered as threat to the entire globe. Much narrower and parochial geo-political interests of the dominant US and its allies were justified in the name of global interests. Much of the policy literatures and debates linking environmental issues and security have been oblivious of and impervious to the danger of what so far the traditional ethnocentric and geopolitical assumptious in US and its allies' security thinking, have defined the world order and global security.[7] This means, understanding of new concerns about environment within the traditional framework of security marks a deflection from what the original proponents of environmental security have had in their mind. If the direction of tide of debates on security concerns are any indication, then in recent years security thinking has become much more importantly part of the process of international politics and the formation of US foreign policy.[8]

This chapter makes out a strong case aiming at fundamentally rethinking the entire ambit of security as a strategy and policy, as well overhauling the ethnocentrism of western political and economic metaphysis. Against this backdrop, while launching a broadside at the geopolitical edifice of the environmental security, what it imperatively adduces a fundamental critique of the world under as fashioned by the US and its allies' political and economic thinking . Unless and until in the new situations created by the crumbling of cold war, environmental security is not understood in these new ways, it is quite possible that environmental security " may turn out to be just another chapter adding to the pages of traditional cold war security concerns of US and its allies with maintaining the global political and economic status quo.[9]

Environmental Security : Not a Security to US Geopolitical Order, Social System and Way of Life

No sooner had the second world war ended than the United States, the major and dominant player started pontificating to the world its own version of what world political and economic order would usher in . The language and idioms the United States used to spell out its own script of world political and economic order seemed to be largely ingrained in its own domestic and capitalist political and economic structure which has been built by the consent of its people for a liberal market based polity and economy. This liberal market oriented polity and economy founded on Hobbes and Locke's possessive individualism aimed at ensuring abundant economic growth, wealth production and providing a luxurious, materialistic and commdious living to its people. On this metaphysical anvil US shaped its world order and foreign policy behaviour. For this the United states embarked upon massive military build up and nuclear weaponisation which would enable it to have unimpeded access to resources and to counterveil the outside forces in general and the geographical expansion of Soviet influence in particular. These geopolitical strategies of US built on the foundation of US domestic imperatives of providing abundant wealth production and inordinate life style to its people deterred outside challenges to its shaping of what political and economic order was best conducive to its national interest. This meant dividing up the world and containment of Soviet Union.[10] This particularistic and domestic imperatives of US resulted in domesticisation of the entire globe and subjecting it to US beck and call. The military development and economic assistance, military alliances and foreign interventions resulted in some degree of control of global economy by the US to ensure the flow of key resources to the United States and its allied powers like Europe and Japan.[11]

So long as the western pattern of development is based on extraction of non renewable resources by fossil fuel technology, the flow of resources unimpededly from the South would remain key to the geo-political understanding of security prevalent during the cold war. Any discourse on environmental security can thus be explicated as merely an extension of conventional security to have the continuance of the resource flow from the South to the US and its allies. Whether it is at Rio, Kyoto, Johannesburg or other international forums on environment, the blame is not put on the US and its allies-the villian of the piece who have caused nuisance to the environment by their own pattern of development and inordinate life style. The culprit admonishes the South and the poor to slow down population growth, to curtail dependence on use of natural resources and suspend economic activities and development process. To the Southern critique, these are mere political gimmicks on the part of those in North who find in the maintenance of their control over global politics and resources, the assurance of security to their sovereign state system and their people against external challenges to wealth production, economic growth and inordinate life style. The nuances of environmental security not throwing focus on this conventional understanding of security appear to reinforce what the US and its allies have considered as fundamental to define their geopolitical order and global security on their own idioms.

It is alleged, the North is using fears of environmental disasters as a weapon to limit Southern development, or to control its political arrangements to ensure that Northern economic interests continue to have access to the traditional cheap flow of labour and resources.[12] Thus, environmental security is understood as securance of the environment to the North - the resources of the nature from the South countries at their perils. Questions confronting any discussion of the themes of environmental security as raised by Southern critique remain unanswered what is being rendered secure, by what means and for whom and what the threat is to this security. In the cold war and conventional parlance, what is threat to the security of the environment is not the western pattern of development, its political and economic thinking and luxurious life style, but the outside or external powers or states challenging the dominant role of the US and its allies in defining the global geopolitical order and security. Unless and until this Western determined geopolitical security thinking surviving this day even after the eclipse of cold war is rethought, attaching the term "environment" to security will be a misnomer conveying nothing substantial and meaningful different from what is conventionally understood as security.

Protection of the national sovereignty and territorial integrity of a state system is what gives birth to security in the conventional sense. In its logic it is always understood in geopolitical divisions of inşides or outside, internal US and the external other.[13] During the cold war it is understood as spatial distancing of others and deterring and counteracting the external threats emanating from other states by recoursing to military forces. Here the state remains importantly as the referent point around which all domestic political resources are mobilized to face the external threats. It means the externalization of the ethnocentric images of state as the sole legitimate authority in the form of constellation of all its military forces to not only define what constitutes the threat but also counterveil it. This conventional security is based on the assumption that the outsiders, or external others are always threats to the internal us. Thus, the externalisation of its internalized domestic and political forces in the form of military has always been reactive and passionately responding. Thus, in the traditional definition of security, the state remains the main focal and referrent point. But this state centric ascent of traditional concept of security is supposedly to be absent in the definition of what constitutes the environmental security. Since the focus is on environmental scarcity or degradation of natural resources, and its social effects in terms of poverty, economic decline, population displacement and growing incapability of state, the security threats are certainly noh military as understood by the concept 'environmental security'. Arguing for a military response by the state to these non-military threats means juxtaposing the traditional concept of security to the environmental security in the new situation of the global environmental crisis. Thus, tagging on the environmental security again to the state as the main referent point to face the non-military threats, means denuding the former of its foundation. The danger with this tagging on to the traditional state centric security is that the response to environmental problems will be militarized.[14] Critics often cite Homer Dixon's study of the linkage between environmental degradation and violence in various part of the globe, to justify their point.[15] What is not understood in Dixon's analysis of the main thrust is that the environmental scarcity or degradation of resources coupled with its social effects can even lead to situations response to which has often been violent. This is not an apologetic of the violent or military response but wary of the likelihood of military and violent response, if environment is not put in order. That is what he tries to drive home the problem, which the critics and analysts do associate with the definition of environmental security. For example, the migration caused by environmental scarcity of land in Bangladesh to Indian states has resulted in violent group identity conflict between the local people and migrants in recent years with some involvement of both India and Bangladesh.[16] Rather than dealing with the causes and problems of environmental crisis, the response has been to its consequences within the traditional framework of state. The geo-political determined security of the western powers headed by the US was in the context of a legitimacy they sought from their respective people on the grounds that they would follow the liberal market oriented polity and economy strictly to its nuances, which means pledging abundant economic growth, wealth production and an exorbitant life style to their people. For the achievement of this goal the military security apparatus of these western states were geared to secure the procurement of resources from the South at a price the former wanted even at the cost of the latter's economy, survival and environment. When the consequences of western power's adherence to this politico-economic logic have been the catastrophic destruction of the environment in terms of climate change, ozone hole, desertification, global warming, sea level rise, bringing the environmental security to the center-stage of policy making, how that western politico-economic logic determining the traditional frame work of security would again be the foundation of environmental security? No where this paradox is seen more visibly than in international forum on environment where the western powers led by US try very successfully to disown any historical responsibility for what so far has been done to the environment by their politico–economic logic. Further, to stress on such a logic in a different and changed context to be the guiding nuances of environmental security means to divert away attention from internal matters of inordinate consumption, luxurious life style and resource usage in the rich industrialised North where a small minority of world's population consumes a disproportionate amount of world's resources not only producing large quantities of pollution and green house gases but also subjecting the majority poor in South vulnerably to the vicissitudes of nature. According to the 1994 World Development Report, during the period from 1960 to 1991 the ratio of the percentage of the world's wealth held by the richest 20 per cent of the World's population to that by the poorest 20 per cent spiralled from 30:1 to 61:1.[17] Any kind of pointer to the internal matters of consumption and life style is considered as a threat to the rich industrialized North headed by US. The very basis of environmental security should have been to rethink the internal consumption life style and pattern of development of western powers. To the contrary, for example on climate change talks, negotiations and discussions, the focus is not on rethinking the western consumption and politico-economic determined military security and on owning the historical responsibilities rather on dragging the victims – the developing countries to clear the mess for which they are not responsible. The United States now reneged on its commitment to reduce emission of CO_2 by 7.5 per cent agreed to at Kyoto, on the ground that it would affect its economy and life style of its people. That means, any plan or efforts to define environmental security as rethinking the politico – economic logic or metaphysics of the traditional security of western powers comes a cropper with the latter not prepared at all to cause any economic hardship to its people and seek legitimacy from its people on grounds other than promoting economic growth and luxurious life style at the cost of nature and the other people in the South. Thus, the problem to the North is any kind of threat to its internal consumption and focus is divagated to the possible political disruption from environmental refugees and migrants caused by external environmental degradation. In stead of looking to the internal causes, the attention of the North is glued to the consequences of external environmental degradation as threats to its traditionally understood military security apparatus.

Violent conflicts, political disruption and turmoils as found in Homer Dixon's analysis, for example in Africa, are portrayed by the western powers as independent and unrelated to their control and dominance of global economy and pattern of living promoting undevelopment and poverty.[18] Environmental security understood in these traditional terms has been justified as a part of wider design of the western powers led by US to secure the resources to its people and economy, manage the environment and control the international environment ,and reinforce their geo political logic even after the collapse of the cold war.[19]

The security of the US and its allies during the cold war was thus to maintain their supremacy in the world economic structure and to stall any kind of external disruptions of supplies of raw materials.[20] The most recent example is the role of the US in the Gulf during the Iran – Iraq war in the 1980s and American attack on Iraq in the 1990s and at present in attempting to secure access to supplies of oil against Iranian or Iraqi threats. The geopolitical strategic security understanding of US was based on having arrogated to itself the exclusive right to these supplies of resources from any part of the globe and the right to militarily intervene for the exercise of that right.

Persistence of focus always by US and its western allies on reducing vulnerability of their political –economy, industrial production, luxurious life style of their people to the changing international situation, say oil crisis or environmental crisis, is what constitutes security as mere external response in military terms to the changing international circumstances and not internal changes or rethinking their political-economic logic. Without bringing changes in their political-economic metaphysics to rest attention on the environmental changes in the South is what the US and its allies do to define environmental security for reinforcement of their economic-political logic and superimposition of their own view of world order on the entire globe as universal one. It was true as much during cold war as now with the end of cold war. This kind of persistence of constancy in North's economic-political thinking despite changes in the environment solely the result of its thinking is what makes environmental security, just an appendage to western conventional understanding of national security.

There is no gainsaying the fact that it is a select few who manage the world's political-economy to their exclusive advantages. If environmental security means forestalling deforestation in the South, making its forests as carbon sinks for the industries of North, and stressing on South's meaningful participation to obviate the misdoing of the North to the environment and close down its industries even for survival, then the North wants to supposedly give protection to its economic-political thinking, way of life and pattern of development what can be generally christened as its civilized life, or modernity or social system. This is the crux of the matter while defining environmental security. What is missing is if the environment is to be secured, preserved and protected, then the Northern pattern of development, way of life, and pattern of consumption are to be modified. In other words, environmental security is stressed on or added to conventional understanding of security in order to give protection to North's view of world order shaped on the anvil of its domestic imperatives and way of life.

Dilemmas of Environmental Security : Problems of Equity and Historical Responsibilities

The questions of global equity and issues of intergenerational responsibilities remain crucial to environmental security and the pertinent question of who and exactly what are to be secured remains tied with environmental security. Every human being contributes to the carbon dioxide concentration in the atmosphere, though the person's lifestyle decides the amount that is emitted. The more prosperous a country's economy, and the higher its per capita income, the higher its fossil fuel consumption for power generation and transport, and the higher its Green House Gases (GHG) emissions.[21] Southern critiques point out that industrialized countries owe their current prosperity to years of historical emissions, which have accumulated in the atmosphere since the start of industrial revolution, and also to a high level of current emissions. Developing countries, meanwhile have only recently set out on the path of industrialization, and their per capita emissions are still comparatively low. The GHG emissions of one US citizens, for instance, were equal to 19 Indians, 30 Pakistanis, 17 Maldivians, 19 Srilankans, 107 Bangladeshis, 134 Bhutanese or 269 Nepalis in 1996.[22] Furthermore the argument runs, as pointed the Southern critique, Southern emissions, and the contribution of tropical deforestation to carbon dioxide and methane production, are often a result of survival activities rather than the luxury emissions that come from Northern consumer lifestyle.[23] Methane produced from a peasant plot is much more immediately necessary to human welfare than an equivalent discharge from an industrial unit in a Northern state producing some luxury consumer item. Whether it is deforestation in Amazan or more gas production in Algeria emitting a large amount of methane, all these are only to meet the Northern demands. Hence the South objects to being asked to pay the price for fixing the problem that North defines as global. Terming the global environmental degradation as of common interest requiring global solution, the North is skirting its own historic responsibility and shifting it to the South for the protection of former's economic interests. This obscures the specific environmental insecurities that the people in South face due to the industrial and economic activities of the North. Again environmental security provides a scapegoat to the North to jettison its responsibility in the name of common interests and global solution. The environmental security, thus remains saddled with rich countries unwilling to compromise their life styles and poor countries unwilling to accept a premature cap on their right to basic development. The atmosphere is a common property resource, to which every human being has an equal right. The people of industrialized countries have more than used up their share of the absorptive capacity of the atmosphere, through their high-emission levels in the past and in the present. So it is only right that they should take the initial responsibilities of reducing emissions while allowing developing countries to achieve a basic level of development. Environmental security if it means asking developing South to reduce carbon emissions levels now, amount to asking them to freeze their standards of living at their current stage of development and freezing global inequality by accepting that some countries will always be more developed than others in the world. Determined not to compromise in any way its consumption pattern, life style and economic development, the North finds in environmental security an extension of its conventional security to control the South through the use of term global concerns which "create the normal base for green imperialism."[24]

Environmental Security: Not an Extension of North into South's National Sovereignty

Global atmospheric change and ozone layer depletion which are mainly due to North's ethnocentric pattern of development and way of life, have become global in their consequences. No single state owns or has jurisdiction over atmosphere or Antarctica or ocean – a global common property but also few states corner the benefits. But the question crops up? Can the forests in tropical and underdeveloped areas be treated as global common to ensure the survival of global biodiversity as suggested by the North in Rio Earth Summit ?[25] The argument suggests that the precious and unique resources inherent in these forests can benefit commercially and medically the future generations, if administered globally. Besides, forests can serve as carbon sinks for stabilization of climate for the North countries. South's vast biodiversity is widely used in west's agricultural system and biotechnological industries. Ascribing such concepts as free access to biodiversity, global commons or common heritage of mankind to forests of the developing world by the developed North not only redounds to the latter's economic interests but runs directly into the question of the political remit of sovereignty.[26] Industrial countries have mainly put emphasis on measures for in situ conservation, demanding the setting aside of protected areas, especially in tropical forests. The economic reason behind this is that they want to ensure a continued supply of actual and potential raw materials for plant breeding, for new medicines and for the wider biotechnology industry. Many developing countries, on the other hand question why should they forgo development of their own territories simply to guarantee conservation for the benefit of the North. They argue that they have sovereign rights over their biological resources, and that they should restrict access to commercial interests in the North, and develop the resources for their own national benefit. They feel, the pressure for forest protection in the South for the development of the North as threats to development of the former that would enrich the state and provide jobs for the growing population. International environmental agreements that constrain development are then seen as a threat to the national security of the state.[27] Much before the concerns about the consequences of devastation of forests by the North were on the political agenda, the latter became highly wealthy and rich through a complete deforestation of territory. Nowhere such unjust and dubious logic is seen than in North admonishing to South to stop cutting trees even for their own survival. What is more glaringly a gross denial of equity and justice is the exploitation of South's biodiversity as well the genetic knowledge of its indigenous people without paying any compensation to the people and state to which the North owe the resources.[28] International agreements on environment such as Clean Development Mechanism (CDM) as agreed to in Kyoto forbidding the South from hewing down trees in the name of environmental security are meant to make North entirely scot-free in emitting CO_2 as before without a chance of being blamed for causing climate change, and deny the South the right to follow the Northern route to development and prosperity. "A Southern response using the same logic might suggest that the American green belt be made a global common resource under an international regime for meeting the improved nutrition needs of the South. Such a proposal would no doubt receive disapproval from the US people. The analogy is imperfect, but heuristic exercise is useful in exposing the North's conventional environment thinking to its limitations, dubious logic and gross injustice.[29]

The structure of the contemporary global economy founded by the North is a very important cause of environmental degradation. Development defined by the domestic imperatives of wealth production, luxurious life style and exorbitant consumption of the North has been universalized as a model of development. Extracting resources from the South to meet the demands of the North has resulted in the shadow economy of degradation beyond the boundaries of the states in which ultimately these resources are consumed.[30] There has been no end to the damage wrought to environment by the conventional Northern modes of development.[31] Dams flood fertile valleys to provide hydro electricity to distant urban dwellers. Forests are logged for the benefit of the urban based industries many of them foreign owned. Rich and precious biological diversities are destroyed. Green revolution may have increased the overall production in specific states that are wealthy enough to bear the cost of expensive chemicals, seeds and equipment. Its impact in terms of soil erosion, fertility loss, pollution of water and diseases has been disastrous. The agricultural export model of development promoted by North and initiated by the South in order to provide foreign exchange to pay for international debts has aggravated the situation. With large enclosure of land for commercial production, the poor move on to marginal lands and steep slops only to destroy the environment. The environmental deterioration and decline of the economy have fed on one another. Ironically the fate of the poor and the fate of the planet have become inextricably intertwined. There is something fundamentally irony in treating the earth as if it were a business in liquidation.

The fundamental question arises, is the environmental security understood in western economic-political idioms meant for the flow of resources from the South to the North at the cost of the former's environment and economy? If security is reformulated to give priority to the environment, it also requires in the same vein to rethink the world economic structure and philosophy of life of the North based on luxurious life style and profligate consumption of resources.

Post Cold War American Perspective of Environmental Security

A detailed glance over the points raised above, suggests that the world economic and political order is the reinforced manifestation of North's political economy. The world order during the cold war or during its eclipse has never been based on democratization of the world economy and trading system, equity in participation and access to resources, equality of sovereign states, collective security, nuclear free world or sweeping powers to United Nations. The ephermeral euphoria about a new world order associated with the demise of cold war fades away into a religious acquiescence in things as they are or still worse into a slavish adulation of the North versed world order. Internet or PC may have acquired death of distance or end of geography but it enables the North to have a unilinear occupation of space and time in terms of universalizing its model of development and initiating others into its economic-political thinking. The entire globe has been at the behest of the North to serve its economic and political interests. Geo political foundation of North's security is based on securing the resources from the South at the cost of latter's economy and environment. Securing these resources has been North's right, realization of which is defined as masculine exercise of military power. The Berlin wall has been tumbled down, but the permanent geopolitical walls in terms of North and South, West and the rest have been reinforced. North's understanding of the post-cold war world security is the persistence of Northern institutions and their core political-economic values as the podium from which the world can be managed. The logic of such thinking is clear in a speech given by Manfred Worner, then NATO's Secretary General :

> The immense conflict potential that is building up in Third World countries, characterized by growing wealth differentials, an exploding demography, climate shifts and the prospect of environmental disaster, combined with the resource conflicts of the future, cannot be leftout of our security calculations, no matter how we translate our broader analysis in operational aspects in the longer term.[32]

According to Dalby, the traditional geopolitical notions of security emphasized territorial control, superior alliance systems, and technological superiority in military matters. In so far these geopolitical themes dominate discussion and debate about Us global political roles, coupling these themes to environmental security is likely to perpetuate patterns of violence and domination rather than work to protect environments and their peoples[33]. Military action and intervention may be justified in seeking compliance from rouge CFC plants. The economic and political metaphysics of the North contributing to its geopolitical security requires to be reconsidered. Fundamentally rethinking international politics, the role of militarisation, externalization of attention to outside changes, global economy and the presupposition of a limitless nature to be exploited, will constitute the parameters on which environmental security is to be defined. The traditional and realist self help paradigm of security is to be replaced by a cooperative, mutuality and consensual acceptance of a diminished national sovereignty. Fundamental and crucial to environmental security is the requirement of readiness on the part of the rich North people to restricting their life style and exorbitant consumption. There needs in the North a reordering of the North's domestic imperatives such as wealth production, economic growth and life style which have contributed to the foundation of its geopolitical understanding of security.

Economic Rethinking : The Basis of Environmental Security

Other challenges to environmental security remain important. Economic rethinking is essential to making environmental security a future possibility. Economy has dominated the entire societal space. The laws of the economy try to dominate the society and not the rules of the society, the economy. The term market economy reduces all the societies or the entire globe to mere an economy ignoring the basic think that every society has an economy. Economy has assumed a deterministic character imprinting its rules and rhythms upon the society and even dictating the drama and role in it of individuals, This "economy is every think and all" approach is to be reviewed so that other considerations and values like frugality, selflessness, moderation in consumption and a humanistic concern for security remain equally important to rebuilding economic structure of the world.[34]

Not universalisation of ethnocentric western model of development, and imitating an other as the cause of development for others but supercession of political categorization like North, South, West and rest and acceptance of diversity, coexistence and holistic cooperation among all states, in transcendence of territories of the world and exploring pluralistic and alternative ways to western model of development and thinking are necessary to define environmental security in terms beyond the geopolitical domination of the west. In the absence of the rethinking of fundamental political and economic metaphysis of west and its associated institutions, arrangements and international politics, the prevalence of liberal market economy and faith of all including the South in the market driven change as a mark of moderity and civilised way of living, makes the prospects of environmental security a remote possibility. The carving of the entire planet into a patchwork of mutually exclusive territorial states is modernity's rendering of natutre.[35] Not all territorial sovereign states synchronize with ecogeographical regions. This asymmetrical incongruity between ethnic, religious, territorial states and ecological units accounts for many of the disputes and conflicts between and among states over the issue of environmental scarcity of water, fish or other resources.[36] Since environmental problems are transboundary and do not end with territorical exclusiveness, the environmental security suggests rethinking of political categories like state sovereignty, and territorial exclusiveness to solve and manage the grave problems of environment.

The global climate change science presents a challenge to a key source of state's sovereign authority its role as guarantor of wealth production and inordinate life style and consumption. Given the close correlation between wealth production and fossil fuel consumption, explanations of the climate change based upon North's traditional economic interests cannot account for the general willingness of industrialized countries to commit themselves to reducing greenhouse gas emissions. That willingness even for the most reluctant countries like the US, can only be explained on the basis of climate change science's certainty of predictions about the disastrous consequences of following the western pattern of development based on fossil fuel technology. Hence environmental security makes the western economic and political logic redundant in view of what climate change science predicts.

Conclusion

As seen from the above analysis the US understanding of security is based on traditional realistic framework. It is wont to defining environmental security in terms of its unimpeded access to resources available anywhere in the world, even without hesitation to exercise military force against the recalcitrant states unwilling to toe the American line. It tends to put the blame on the South countries for what has been done to the environment, so also the responsibility on them to clear the mess for the rich North countries. The North countries headed by US define environmental security in terms of its being just an appendage to their conventional geopolitical security. These countries have underrated or downplayed their own nuisance to the environment in terms of climate change, global warming and ozone layer depletion. The US and its allies' inordinate consumption, exorbitant and profligate life style and production of wealth remain an important basis of their concept of security. The entire planetary resources are to be exploited to the promotion of these interests of US and its allies. The ubiquitous acceptance of western pattern of development, liberal market economy and economic thinking has been an extension of American geopolitical understanding of security. What has been at the root of the environmental crisis is not the burgeoning population growth in the South, rather the domestic imperatives of the North in terms of their life style, pattern of living and limitless consumption.

In order to ensure environmental security to the entire humanity, what the policy makers need is to keep in their mind the following parameters

- The future generations - living non-human and human beings – rights to environment has to be included within the term environmental security. Inter generational accountability and equity is to be given primacy. The very arrogation of sovereign state system to exploit the resources at the cost of others and future generation is to be substituted by the principles of equity and accountability to the future generations.

- While defining development it is not the blind imitation of western pattern but the entire interests of the states, humanity, future generations and living and non living are to be weaved into the frame work of what is meant by environmental security.

- The inordinate consumption, profligate life style and production of wealth of the North countries headed by US are to be rethought so that environmental security can be ensured to all, otherwise environmental security as understood by US will remain an addition to its conventional understanding of security. So the over consumption on the part of the North is normally and ethically indefensible. What remains pivotal to defining environmental security in a broader sense that will remove the causes of environmental conflict as well the causes of environmental scarcity that inflicts on the individuals in terms of health hazards, suffering, economic deprivation, poverty or starvation as undergone by them as a part of adaptation. Thus, at the confluence of the morally and ethically required nuance of sustainable development, economic rethinking, new planetary principles and new vision of and approach to life, stands the environmental security. Thus, the very basis of environmental security challenges the established frame of mind about what so far has been defined as development, pattern of life style, and geopolitical security of the sovereign state system. It is not the national security in terms of military but the planetary security that will be the foundation of environmental security on this confluence.

- The very definition of environmental security requires new modes of thinking that rest on a new – found compassion for the integrity of the planet and for one another. It needs a recasting of human responsibility for planetary stewardship. The notion that Earth must be seen as a living organism – in the sense of a homeostatic equilibrium that demands respect from humanity and other terrestrial life-forms-has a spiritual dimension on what was initially conceived as a strictly scientific hypothesis and external mechanic structure to be exploited to satisfy the inordinate lifestyle of the North people. A comparable rethinking of the human-earth-cosmos relationship is necessary to found the very basis of environmental security.

References

1. Norman Myers."Environment and Security, "Foreign Policy no.47,1989,PP.23-41;Jessica Tuchman Mathews. "Redefining Security". Foreign Affair Vol. 68, no-2, 1989, pp. 162-177; and Michael Renner. National Security: the Economic and Environmental Dimensions (Washington, D. C. : World watch Paper 89, 1989).

2. Thomas F. Homer Dixon, Environment Scarcity and Violence (Princeton New Jersy : Princeton University Press, 1999)

3. Simon Dalby, "Threats from the South? Geopolitics, Equity, and Environmental security", in Daniel H Deudney and Richard A. Matthew ed., Contested Grounds : Security and Conflict in the New Environmental Politics (New York : State University of New York . 1999), p. 156.

4. Commission on Global Governance, Our Global Neighbourhood (Oxford : Oxford University Press, 1995).

5. On Clinton administration policy, see Geoffrey D. Dabelko and P. J. Simmons, "Environment and Security : Core Ideas and U. S. Government Initiatives", SAIS Review Vol.17, no. I, 1997, pp. 127-146

6. Simon Dalby, "Threats from the South ?" *op. cit.*, p. 157.

7. See Robert Young, White Mythologies : Writing History and the West (London : Routledge, 1990), .Ken Booth, Strategy and Ethnocentrism (London : Croom Helm,1979)

8. David Campbell,Writing Security : American Foreign Policy and the Politics of Identity (Minneapolis : University of Minnesota Press, 1992). Cynthia Weber, Simulating Sovereignty Intervention, the State and Symbolic Exchange (Cambridge, U.K. : Cambridge University Press, 1995).

9. See Simon Dalby, " Contesting an Essential Concept : Reading the Dilemmas in Contemporary Security Discource" , in Keith Krause and Michael Williams, ed., Critical Security Studies : Concept and Cases (Minneapolis : University of Minnesota Press, 1997), pp. 3-31.

10. Bradley S. Klein, Strategic Studies and World order : The Global Politics of Deterrence (Cambridge, U.K. : Cambridge University Press, 1994).

11. Ronnie D. Lipschutz and John P. Holdren : Crossing Borders : Resources Flows, the Global Environment , and International Security ", Bulletin of Peace Proposals Vol. 21, no. 2, 1990, pp. 121-133.

12. Alvaro Soto, "The Global Environment: A Southern Perspective", International Journal Vol. 47, no. 4, 1992, pp. 679-705; N. Middleton, P. O'Keefe and S. Moyo, Tears of the Crocodile: From Rio to Reality in the Developing world (London: Pluto, 1993).

13. Simon Dalby, Creating the Second Cold War : The Discourse of Politics (London: Pinter, and New York : Guilford, 1990) See also R. B. J. Walker, Inside / Outside : International Relations as Political Theory (Cambridge, U. K. : Cambridge University Press, 1993).

14. Daniel Deudney, "The Mirage of Ecowar : The Weak Relationship among Global Environmental Change, National Security and Interstate Violence in I. H. Rowlands and M. Greene, eds., Global Environmental Change and International Relations (London : Macmillan, 1992); Lothar Brock, " Security through Defending the Environment : An Illusion ?" in E. Boulding, ed.,

New Agendas for Peace Research : Conflict and Security Reexamined (Boulder, Colo. : Lynne Rienner, 1992); and Jyrki Kakonen, ed., Green Security or Militarised Environment (Aldershot : Dartmouth, 1994).

15. Thomes F. Homer Dixon, Environment, Scarcity and Violence, *op. cit.*

16. N.Gaan, Environmental Degradation and Conflict : The Case of Bangladesh India (New Delhi : South Asia, 1998).

17. United Nations, World Develop Report 1994, p. 35.

18. Robert D. Kaplan, "The Coming Anachy", The Atlantic Monthly Vol.273, no.2, 1994, pp. 44-76. For a detailed critique, See Simon Dalby, "The Environment as Geopolitical Threat : Reading Robert Kalplan's Coming Anarchy", Ecumene Vol. 3, no. 4 (1996), pp. 472-96.

19. Wolfgang Sachs, " Global Ecology and the Shadow of Development", in Wolfgang Sachs, ed., Global Ecology: A New Arena of Political Conflict (London: Zed, 1993).

20. Beverley Crawford, "The New Security Dilemma under International Economic Independence", Millennium, vol.23, no.1, 1994, pp.25-55.

21. T. A. Boden et al, Emissions of Global Regional and National Annual CO_2 Emissions from Fossil- Fuel Burning, Hydraulic Cement production, and Gas Flaring : 1950-1992, Oak Ridge National Laboratory, Oak Ridge : Tennessee, 1995 See also UNDP, Human Development Report 1998 (Oxford : Oxford University Press, 1998) , pp.140-42.

22. Anil Agarwal, Sunita Narain and Anju Sharma, eds., Global Environmental Negotitaions I : Green Politics (New Delhi : Centre for Science and Environment , 1999), pp. 17-18.

23. Anil Agarwal, Sunita Narain ed. Global Warming in an Unequal World : A Case of Environmental Colonialism (New Delhi : Centre for Science and Environment, 1991)

24. Vandana Shiva, "Conflicts of Global Ecology : Environmental Activism in a Period of Global Reach". Alternative Vol.19, no.2, 1994, pp. 195-207.

25. Rachael M. McCleary, "The International Community's Claim to Rights in Brazilian Amazonia" Political Studies Vol. 39, no. 4, 1991, pp. 691-707.

26. Ken Conca, " Rethinking the Ecology Sovereignty Debate", Millennium Vol.23, no.3, 1994, pp.701-11; See also Andrew Hurrell, "A Crisis of Ecological Viability? Global Environmental Change and the Nation State", Political Studies Vol.42, 1994, pp. 146-65.

27. Bruce Albert, " Indian Lands, Environmental Policy and Military Geo-Politics in the Development of the Brazilian Amazon : The Case of the Yanomami", Development and Change Vol. 23, no. 1, 1992, pp. 35-70.

28. Franz Broswimmer, "Botanical Imperialism: The Stewardship of Plant Genetic Resources in the Third World", Critical Sociology Vol. 18, no. 1, 1991, pp. 3-17.

29. See Simon Dalby, " Threats from the South ?" op.cit. p. 169.

30. Jim MacNeill, Peter Winsemius, and Taizo Yakushiji, Beyond Interdependence: The Meshing of World's Economy and the Earth's Ecology (Oxford : Oxford University News, 1991).

31. Bruce Rich, Mortgaging the Earth : The World Bank, Environmental Impoverishment and the Crisis of Development (Boston : Beacon Press. 1994).

32. Manfred Worner, "Global Security: The Challenge to NATO", in E. Grove ed., Global Security: North American, European and Japanese Interdependence in the 1990s (London: Brassey's 1991), pp. 103-104.

33. Simon Dalby, Threats from the South ? *op.cit*, p. 175.

34. Wolfgang Sachs, " Development : a guide to the ruins", New Internationalist, June 1992, p. 5

35. Karen T. Litfin, "Environment, Wealth, And Authority : Global Climate Change and Emerging Modes of legitimation", International Studies Review special issue Vol. 2 issue no.2, Summer 2000, p. 127.

36. Arthur H.Westing ed., Comprehensive Security for the Baltic : An Environmental Approach (London : Sage publication, 1989), pp.1-13. See also N. Gaan, "Comprehensive Security for South Asia : An Environmental Approach", BIISS Vol. 20, no. 2, 1999, pp. 103-105.

8

Global Climate Change : Challenges to Sovereign State System and Its Legitimacy

The modern sovereign state owes its origin to the Treaty of Westphalia that was signed in 1648. The idea of territorial exclusivily, central to the sovereign state system, took shape at Westphalia. This implies that each state has the absolute right to utilize its human resources without any hindrance from any quarter. But the concept of territorial exclusivily is not always in conformity with ecogeographical regions. An ecogeographical region is essentially an ecological subsystem made up of living and non-living components of the environment that interact to form a life support system. It is by no means fully self-contained as it has relationship with subsystems in adjoining areas and beyond. Being a part of the global ecosystem, it has its own dynamics to function independently of the contiguous and more distant regions, as well as of the globe as a whole.[1] Thus, lack of coordination between the sovereign state system and the ecogeographical regions, in the face of environmental degradation, has often resulted in transitional and global environmental problems. Efforts to cope with these problems pose a challenge to the sovereign state system. Sovereign territorial exclusivily presupposes that political authority can address problems arising within its jurisdiction. This becomes problematic in the context of global ecological interdependence. Thus, " territorial exclusivily stands in sharp contrast to the outlook of global ecology, based on holism and the mutual dependence of Earth's ecosystem".[2]

The sovereign state system gradually evolved into a liberal, democratic political order based on the consent of the governed. Premised on the Hobbesian and Lockean concepts of atomistic and possessive individualism and mechanistic worldview, liberal democracy ushered in a political authority, i.e., government –the political legitimation of sovereign state system- to cater to the economic prosperity of the governed and tend to attest the sovereign state system's exclusive right to exploit nature. Based on these notions, the modern sovereign states used nature in a way which resulted ecological consequences in a world of finite resources.

The authority of the modern state derives from its role as the guarantor of territorial integrity and wealth production. On these accounts, many third world and developing state have failed to do the needful and become dependent on external powers causing the erosion of their sovereignty. These are at times referred to as "quasi states" which derive their authority more from their external recognition than from their indigenous capacity to facilitate wealth production.[3] The state's authority rests substantially on the government's ability to protect economic interests which in turn requires greater cooperation in the management of international economic transactions, compromising much of its autonomy, independence and exclusiveness.

Traditional Concept of Political Authority in International Relations

Authority means power legitimized. In the Western context, it is understood as the rightful regulation of human behaviour by means other than coercion or persuasion.[4] The question "what constitutes rightfulness?" has always been answered in terms of consent of the governed that is institutionalized through a democratic process. Legitimacy-key to the modern notion of authority –depends upon the extent to which "those addressed by a rule understand themselves as being obliged by it rather than merely coerced by whatever power created and supported it."[5] In other words, the individuals assume the character of reflective citizens quite capable of judging whether the supposed actions, rules, or proposals are in furtherance of their interests or in accordance with established norms and practices. This becomes a problem not in the domestic context but in the international realm.

The international relation theorists adopt either a juridical interpretation of authority or associate authority with power. For those who adopt the former approach, legitimacy is found in the international recognition by other states or in the fulfillment of requirements of international law.[6] While giving primacy to the normative character of legitimacy, they emphasize inter-state rather than state-society relations. Thus the juridical concept of authority has reduced legitimacy to the "collective judgement of international society about rightful membership in the family of nations."[7] This does not take into account the people constituting the state. On the other hand, those realists who conflate authority with power or autonomy find its legitimacy in its ability to maintain the territorial integrity of the state. In both approaches, what remains distinct is the disappearance of people as they seek to legitimize authority by referring to behaviour or perception of other sovereign states. In international relations states never act or judge, it is the governments, policy makers and leaders who are the real actors. However, their fate remains tied to the consent of the governed. The idea of consent remains open to various interpretations. Critical theorists contend that "the notion of consent tends to equalize the position of subjects and obscures the asymmetry of power relations "between those who govern and the governed.[8] It does not take into account the disempowered section(s) of the society which may not have given its/their consent. Nevertheless, it remains crucial to the idea of legitimate political authority, for the aim of theorists and critics alike is "to develop a doctrine of political authority which makes its legitimacy conditional on the existence of a population which regards the governments as its representative." [9]

The concept of popular sovereignty assumes importance in domestic politics; in international politics, it is almost non-existent. But with the dramatic increase in transnational activities of citizens belonging to different countries and of non-states actors over the past two decades, the necessity of popular participation at the international level has begun to be felt.[10] There can be no ground for confining democratic practices to the territorial limits of a nation-state only.[11] Transnational environmental activism, movement for globalization and the gradual emergence of a global civil society highlight not only the importance of non-state actors but also the emergence of new modes of legitimation and new types of authorities. The political authority representing the sovereign state system enjoys exclusive jurisdiction over exploiting nature within the confines of its territory. While doing this, the political authority does not take into consideration the interests and rights of the future generations. The intergenerational responsibility which includes the future generations' right to a healthy environment, access to fresh and unpolluted water, air, and other resources and the present generation's rights and obligations provides sufficient grounds for reviewing the sovereign state system. The crucial issue is how can the future generations give their consent to the present political authority's jurisdiction over the exploitation of natural resources ? Since environmental problems have become global concerns, transnational environmentalism "aims to transform the modern conception of authority to include not only popular legitimacy, which has emerged in tandem with spread of democracy, but also inter-generational and even interspecies criteria of accountability, based on intergenerational equity."[12]

Tackling environmental problems such as acid rain, pollution of air and oceans, ozone layer depletion, global warming and climate change involve non-traditional forms of external inventions which challenges the sovereign state's omnipotent authority to control events within its territorial jurisdiction. The very raison d'etre of the sovereign state system to protect its people and keep its territorial exclusivity intact comes under attack from these non-military sources. The link between environmental degradation and national security has recently become the major concern of sovereign state system. Environmental security goes much beyond the traditional realist paradigm of security. It includes watersheds, croplands, forests, genetic resources, climate and other aspects of nature so that "the exploitation of renewable resources must be carried out strictly on the principle of sustained use or sustained discard, with exploitation of non-renewable resources strictly on the principle of frugality."[13] It requires self-imposed restrictions on sovereign states' right to exploit nature to satisfy the greed and avarice of its citizens.

Tackling the problem of environmental degradation may require new strategies, which may be beyond the capacity of many third world and developing countries. Therefore, international environmental responses may "involve trade offs among autonomy, control and authority".[14] In other words, the imperatives of environmental degradation may require all nations to compromise their unrestricted freedom in relations to their natural resources, industrial practices and the environmental.[15] Contrary to claims made by some, this requires the concept of merged sovereignty which comes into existence when nations seek to realize their own national objectives along with environmental security through cooperation, voluntary remedial measures and owning responsibility in association with other nations of the world keeping in view intergenerational equity and accountability. Thus, the traditional view of sovereignty seems to be inconsistent with the requirements of environmental protection.[16]

Interplay Between Political Authority and Authority of Science

Economics performance and a consumerist pattern of life have been linked to the exponential growth in fossil fuel production. The western pattern of development based on industrialization and fossil fuel production has become a standard model for both capitalist and socialist states. As a result, large quantities of carbon dioxide are emitted to the atmosphere causing pollution and an alarming climate change. A majority of scientists believe that this is the primary greenhouse gas (GHG) responsible for global climate change which seriously limit the state authority's capability to protect its citizens. In spite of efforts to find alternative energy resources, the authority of science poses a challenge to the authority of states as the sole determinant of wealth production. The signing of the 1997 Kyoto Protocol by the industrialized nations committing themselves to a reduction of GHG emissions by 7 to 8 per cent by 2012 is a clear recognition of the authority of climate science. More than any other international environmental issue, global climate change envisages the advent of new patterns of authority based on science impacting on the authority of the sovereign nation-state.

One of the essential characteristics of the modernity is its faith in the role of science to improve the lot of the people. Material achievements and industrial development reinforced this view.[17] It became associated with state when the latter utilized it to improve human life. The rise of modern liberal political theory also coincided with the rise of modern science. The liberal political theory of possessive individualism founded by Thomas Hobbes and John Locke[18] was shaped by the mechanistic world view and Newtonian view of motion.[19] In his "Two Treatises of Government", John Locke argued that the purpose of the state is to protect men's natural rights, particularly the right to property. Consequently, citizens can overthrow a government that fails to protect these rights. Locke's concept of property rights, man and the world are having serious ecological implications. The political authority premised upon his concept of possessive individualism is the guarantor of property rights and facilitator of market relations. Individuals exploit nature to maximize their happiness by accumulating wealth. In a world of finite resources, the modern political authority will at some point have to ensure the protection of the earth and its environment. Thus, science and politics are grounded on the same ontological assumptions. Scientific truths are based on consequences within a scientific community in the sense that any scientist can reach the same conclusion if he/she applies the same method. It is believed that science functions in a "rather rarefied atmosphere immune to the vagaries of political power and subjective opinion, but as in politics, consensus stands as touchtone of authority".[20] There is a possibility that science may be affected by nonscientific forces such as peer pressure, tradition, and desire for funding and socio-psychological identities. While addressing socio-psychological problems, it is likely to enter into the political arena. However, in order to keep its legitimacy it must refrain from political involvement. This dilemma comes to the fore when policy makers depend upon it for identifying and addressing environmental problems.

Another factor which undermines the authority of science is lack of certainty about the scientific truth. In such a situation, the problem of environmental policy making is compounded. Actors and policy makers may fail to comprehend the situation in its totality. This kind of uncertainty is found in global climate change, whose study could be considered a horizon science, not a core science. Horizon science is transdisciplinary in nature. In involves identifying a new problem or new tools to understand an old problem; it may even approach a problem at a wholly different level of abstraction.[21] It is mostly characterized by uncertainty and therefore its credibility in the policy making level remains low.[22] Since the world's climate involves a continually changing and highly complex set of relations, search related to climate change faces a number of obstacles in establishing a linkage between theory and empirical evidence. Climate science relies on elaborate computer models. One such model refers to the General Circulation Models (GCMs). Its accuracy is tested by comparing predictions of past global climate with historical data. Modelled predictions of future climate change are based upon incorporating current and projected GHG emissions into GCMs. Whether or not the modelled predictions are valid cannot be known until sometime in the future, but they currently constitute the most authoritative knowledge available to guide climate change policy.

It should be remembered however that thanks to scientific predictions about climate change, intergenerational responsibility and equity have become political issues. Since the modern notion of political authority, as stated above is based on the consent of the governed, the predictions about climate change question the very legitimacy of the government to deprive the future generations of their right to a clean climate in the name of promoting the economic well being of the present generations. The states which have lagged behind in wealth production and economic well being would try to emulate the fossil fuel based development pattern of the western industrialized nations and resist any attempt at the international level to put restrictions on them. The island states, which are likely to be seriously affected by the rise in sea level due to climate change, may hold the industrialized nations largely responsible for this, impelling them to take concrete steps towards reduction of greenhouse gases. Challenges to the authority of science may also emanate from some known quarters like coal, petroleum and automobile industries and even from states on which these industries have strong influence.

The industrial lobby often relies on scientific predictions, which regard global warming as a myth. It is felt that the US government is most prone to such pressure tactics employed by the industrial lobby. The industrial lobby argues that if the uncertainties are great, so too are the consequences of misjudgement. If it is found that the planet is not experiencing a significant climate change, then the money wasted on research to prevent such a change would run into billions of dollars. The US based Global Climate Coalition (GCC), which was founded in1989 with American Petroleum Institutes, Shell Oil, Exxon, General Motors and more than forty other corporations as members, is the biggest and most powerful of these corporate think tanks. Since 1994, it has been spending more than US$ 1 million annually to downplay the threat of global warming and climate change. The GCC has succeeded not only in influencing the government not to ratify the Kyoto protocol but also funded some scientists to declare that global warming is a myth. The information Council for the Environment (ICE), another US-based corporate coalition, was formed in 1991. Comprising the National Coal Association, the Western Fuels Association and Edison Electrical Institute, the ICE launched a US$ 500,000 advertising campaign with the help of Patrick Michaels (a leading scientist from the University of Virginia's department of environmental science) to propagate that global warming was a myth. In September 1997, the US Automobile Manufacturers and Oil Company launched a US$ 13 million advertisement blitz to convince the public that a strong climate treaty would result in spiralling price rise and would benefit the poor in the South at the cost of the US consumers.[23]

The influence of the industrial groups over the national government, due to scientific uncertainties, is a challenge to the authority of science and has delayed the process of finding a solution to the climate change. But the international institutions are less likely to be plagued by such kind of pressure from the industrial lobby. The Kyoto protocol, which was signed by thirty-nine industrialized nations in December 1997, has been the result of recognition of the authority of science. Once ratified, it will assume the character of a binding treaty and will have far reaching implications for the sovereign states in the 21st century in view of its accountability to the future generations.

Greenhouse Gas Emission and Increase in Gross World Product

It has been seen that between 1950 and 1995, a rapid increase in the consumption of fossil fuels quadrupled the growth in Gross World Product corresponding to a similar growth in the GHG emissions at the international level. Given the fact that GHG emissions produced in the pursuit of prosperity, proposals to reduce these gases have faced strong resistance. A study, conducted in 1988, concluded that stabilizing atmospheric concentrations of carbon dioxide and bringing it back to 1950 levels, would mean reducing fossil fuel use by 50 to 80 per cent.[24] It is estimated that the cost of halving GHG emission to the US alone would be $180 billion.[25]

The Organization of Petroleum Exporting Countries (OPEC) and other states, which are dependent on fossil fuels, oppose reduction in carbon dioxide emission. Given the correlation between wealth production and fossil fuel consumption, the industrialized nations of the North would be willing to reduce greenhouse emissions only on the basis of scientific certainty and accuracy about the climate change. The target set by the Kyoto protocol for the reduction of GHGs in the atmosphere is minimal in view of the dangers the emission of CO_2 by the industrialized North pose to humanity. Even this minimal reduction target has raised eyebrows in some quarters. Critics point out that if implemented, "the treaty would force the most productive societies on earth-the ones that have led the way in making human life comfortable, safe and prosperous—to slow their economic growth and degrade their standard of living".[26] However, the latest report of the UK Meteorological Office's Hadley Center for Climate Prediction and Research reveals that even if GHG emissions are stabilized which would require an overnight cut of 60-70 percent in global carbon dioxide emissions, there would be a 1°C rise in atmospheric temperature and about one meter rise in sea level. Based on the findings of the Intergovernmental Panel on Climate Change (IPCC), the Center projects that warming over land would be 80 per cent faster than over sea and the highest emission scenario could lead to 6°C rise in temperature over land by 2100, 2°C higher than previous estimates.[27]

Climate Change Regimes

It was the scientists who put climate change on the political agenda. Prior to the formation of the IPCC, their conferences in Villach and Bellagio during the 1980s called for a rapid reduction in the use of fossil fuels with the sole objective of limiting the rate of global warming to 0.1°C per decade. At the Toronto Conference on the Changing Atmosphere, over three hundred scientists and policymakers called for reducing carbon dioxide emissions by 20 per cent of the 1988 emission level by 2005. The US EPA, for example, estimated that stabilizing atmospheric concentration of CO_2 at the 1988 level would require reducing greenhouse gas emissions by 50 to 80 per cent. Following the Toronto Conference, the IPCC was established in1988 by the World Meteorological Organization (WMO) and the United Nations Environment Programme (UNEP) to assess available scientific information on climate change, its environmental and socio-economic impact and to formulate strategies for managing it. In1990, the First IPCC Assessment Report was accepted as the scientific basis for international negotiation. Following it,, the UN General Assembly established the International Negotiation Committee for a Framework Convention on Climate Change (FCCC).

The stated objective of the FCCC, which was signed at the 1992 Earth Summit at Rio and came into force in 1994, "is the stabilization of greenhouse gas concentrations in the atmosphere at a level that would prevent dangerous anthropogenic interference with the climate system" (Article 2). The first IPCC assessment had accepted that the greenhouse effect was a reality, that human activities were increasing atmospheric concentrations of GHGs, that mean global temperatures has risen between 0.3°C and 0.6°C during the twentieth century and that the predicted global warming of 1.5°C to 5.0°C would occur sometimes towards the middle of the next century."[28] In its second report of 1995, it concluded that "the balance of evidence, from changes in global mean surface air temperature and from changes in geographical, seasonal and vertical patterns of atmospheric temperature, suggests a discernible human influence on global climate."[29] As regards environmental effects only sea level rise was predictable with certainty. Effects of global warming are noticed at the local and regional levels. Since scientific predictions pertained to global average on long- term basis, the uncertainties about environmental and socio-economic predictions persisted.[30]

The Kyoto protocol mandates reductions on overall GHG emissions by 5.2 per cent of 1990 levels by the year 2012, with most European countries bound by 8 per cent targets, the US by 7 percent and Austria by 8 per cent. It provides for three flexible mechanisms to assist industrialized countries to achieve their targets at a low cost. These are emissions trading (trading of emissions allowances between industrialized countries), joint implementation (industrialized countries can earn credits by financing projects that cut emissions in other industrialized countries) and Clean Development Mechanism (a form of joint implementation between industrialized countries and developing countries). The treaty allows countries to count forests as carbon sinks, so that their overall reductions in actual emissions could be lower than the mandate targets. Even through developing countries are not bound by the treaty, they commit themselves to a Clean Development Mechanism. The future of the treaty remains uncertain as the US has not only refused to ratify the treaty but also coutinues to insist on meaningful participation of the developing countries. The US position is not accepted by China, India and the European Union. Without a major move towards zero carbon energy systems by the industrialized countries and without the acceptance of equity, all negotiations towards reducing carbon dioxide emissions will remain hostage to the interests of the industrialized North.

Conclusion

The very fact that the industrialized nations have signed the Kyoto protocol indicates that the authority of science has made an inroad into the legitimate rights of sovereign nation-states to pursue economic growth on the basis of fossil fuel technology. This implies that the people, on whose consent the sovereign state authority rests, are conscious of the climate hazards and accept their governments' decision leading to a decrease in fossil fuel based production of wealth. It can be argued that the people's willingness to suffer economic loss is indicative of the primacy of authority of science in governance. This shift marks the ascendancy of the authority of science in the process of policymaking by the sovereign state. Therefore, with the emergence of new bases of consent and authority pattern in the climate change science, the traditional sovereign state system undergoes a change.

The kind of international accountability involved in the climate change talks and treaties is suggestive of the fact that the criteria for the consent of future generations of the government is not the fossil based economy but prevalence of a healthy carbon-free climate.

Influence of the industrial groups over the US decision-making process indicates, on the other hand, that people may not be willing to choose any other basis of their consent than the production of wealth based on the fossil fuel technology. In that case, people are not prepared to sacrifice their present economic gains for the future generations. If so, the tension between the authority of science and political authority continues to persist. The uncertainties surrounding scientific predictions may strengthen people's unwillingness to sacrifice their present economic gains and to understand the basis of climate hazards. The more climate science will be certain and accurate about its predictions the sooner will the tension be resolved in favour of the authority of science.

The low target agreed to by the industrialized countries under the Kyoto protocol gives rise to the belief that in the context of the magnitude of the crisis, the agreed upon commitment is very minimal. In view of the increasing pressure from environmental NGOs and international environmental regimes, the industrialized countries may not be able to run away from their commitments to the reduction of carbon dioxide for long. Further there also exists a strong bond between economic prosperity and greenhouse gas emissions, which is difficult to break. For that to happen, science needs to provide alternative sources of energy that would not affect the existing level of economic prosperity and redefine the present development paradigms linking fossil fuel with wealth production.

What has been said above does not in any way reduce the legitimacy of political authority. In the context of global climate change, intergenerational responsibility binds the sovereign state to a new ethical bond and prevents it from pursing policies that have serious consequences for the present and future generations. This implies an awareness of a new image designed by the authority of science. The authority of science which emerged as a solution during the industrial revolution to the problem faced by modern sovereign states has now become a problem itself due to the effects of industrialized and western life pattern on the environment. This is the internal challenge that the authority of science faces at present and is likely to get involved vis-à-vis political authority .The future generations acceptance of the legitimacy of the present sovereign state system can be ensured only when the authority of science comes out with much certainty consensus and finality proving alternatives to the basis of individuals consent to the state in the broader context of the organic bond that exists between nature and the present and future generations, both living and non living.

References

1. L.H.Brown, "Regional Collaboration in Resolving Third World Conflicts", Survival Vol. 28, 1986, pp. 208-20. See also L.J. Cantori and S.L. Spigel.eds., International Politics of Regions : A Comparative Approach (New Jersy : Prentice Hall, 1970), p.432.

2. John G.Ruggie, "Territoriality and Beyond : Problematizing Modernity in International Relations," International Organisation Vol. 47, 1993, pp. 144-73.

3. Robert H.Jackson, Quasi-states : Sovereignty, International Relations, and the Third World (Cambridge : Cambridge University Press, 1990).

4. R.B.Friedman, "On the Concept of Authority in Political Philosophy," in Joseph Raz, ed., Authority (New York: New York University Press, 1990), pp.56-91.

5. Olav Schram Stokke, "Regimes as Governance Systems," in Oran Young, ed., Global Governance: Drawing Insights from the Environmental Experience (Cambridge, MA: MIT Press, 1997) See also, Thomas M.Franck, The Power of Legitimacy Among Nations (Oxford: Oxford University Press, 1990).

6. Hedley Bull and A.Watson, eds., The Expansion of International Society (Oxford; Clarendon Press, 1982). Alan James, Sovereign Statehood: The Basis of International Society (London : Allen and Unwin, 1986).

7. Martin Wight, Systems of States (Leicester : Leicester University Press, 1977).

8. A Clarie Culter, "Locating Authority in the Global Political Economy", International Studies Quarterly Vol. 41, no. 1 (March) 1999, pp. 59-81.

9. Joseph Raz, "Introduction," in Joseph Raj, ed., Authority (New York University Press, 1990), p.17.

10. Margaret Keck and Kathryn Sikkink, Activists Beyond Borders : Advocacy Networks in International Politics (Ithaca : Cornell University Press, 1998).

11. William Connolly, "Democracy and Territoriality," in F.M.Dolan and T.L.Dunn, ed., Rhetorical Republic: Governing Representations in American Politics (Amherst: University of Massachusetts Press, 1993).

12. Karen T.Litfin, "Environment, Wealth and Authority: Global Climate Change and Emerging Modes of Legitimation," International Studies Review (Special Issue) Vol. 2, no. 2 (summer) 2000, p. 127.

13. Arthur H.Westing, "Comprehensive Human Security and Ecological Realities", (Guest Comment) Environment Conservation Vol. 116 (winter) 1989, p..295.

14. Karen T.Litfin, "Sovereignty in World Ecopolitics," Mershon International Studies Review Vol. 41, 1997, pp. 167-204.

15. Lynton Keith Caldwell, International Environmental Policy: Emergence and Dimensions (New Delhi: Affiliated East-West Press, 1991), p.311.

16. Maurice F.Strong, " One Year After Stockholm : An Ecological Approach to Management," Foreign Affairs Vol. 51 (July) 1973, p. 703.

17. Robert Heirman, Authority and Liberal Tradition : From Hobbes to Rorty (New Brunswick : Transaction Publishers, 1994). Yaron Ezrahi, The Descent of Icarus : Science and Transformation of Contemporary Democracy (Cambridge, MA : Harvard University Press, 1990).

18. C.B.Macpherson, The Political Theory of possessive Individualism : Hobbes to Locke (Oxford : Clarendon Press, 1962).

19. Carolyn Merchant, The Death of Nature: Women, Ecology and Scientific Revolution (San Francisco: Harper Collins, 1990).

20. Litfin. "Environments, Wealth and Authority", p. 30.

21. Eugene A. Rosa and Thomas Dietz. " Climate Change and Society: Speculation, Construction and Scientific Investigation," International

Sociology Vol. 13, 1998. See, Stephen Cole, Making Science: Between Nature and Society (Cambridge, MA: Harvard University Press, 1992).

22. Litfin, "Environment, Wealth and Authority," p. 131.

23. Anil Agarwal and Anju Shrama, "A farce of a face off," Down to Earth Vol.6, no. 15, 31 December 1997, pp. 31-31.See also, David L.Levy and Daniel Egan, "Capital Contests: National and Transnational Channels of Corporate Influence on the Climate Change Negotiations," Politics and Society Vol. 26, no. 3, September 1998, pp. 337-61.

24. Christopher Flavin, "Slowing Global Warming", in Lester Brown, ed., State of World 1990 (New York: Norton, 1990), pp. 17-38.

25. William D.Nordhuas, "Greenhouse Economics : Count Before You Leap", The Economist, 7 July 1990,pp.19-22.This also figured in a study conducted by National Academy of Science. National Academy of Science, Policy Implications of Greenhouse Warming (Washington, D.C.: National Academy Press, 1991).

26. Jeff Jacoby, "Scientists Do Not Agree on Global Warming," Boston Globe 5 November 1998, p. 27.

27. Down to Earth, 31 December 2000, p. 27.

28. Intergovernmental Panel on Climate Change (IPCC). First Assessment Report, Geneva: World Meteorological Organization and United Nations Environment Programme, 1990, pp. 2-5.

29. IPCC, 1995, p. 5.

30. James F.Kasting, "The Carbon Cycle, Climate and the Long Term Effects of Fossil Fuel Burning", Consequences Vol. 4, no. 10, 1998, pp. 15-27.

9

Rediscovering New Planetary and Transcendental Vision : Foundation of Environmental Security

Humanity today stands on a precipice of ecological holocaust staring ghastly into the abyss of brink to which it slides irreversibly by the very western concept of development based on fossil fuel technology and mindless industrialization and its ubiquitous practice throughout the world as a testament of progress, and fulfillment of a civilized life. Nothing has come in for greater challenges than the very "development", a western terminological exactitude in the name of which irreparable damage has been wrought to the ecological fabric of our universe. What is more sordid and the crux of the problem is the very suffering from conceptual failure and pretending of the policy makers, leaders and world actors that the crisis does not exist. Pretending is a form of wish fulfillment wherein decision makers adopt the belief that a particular problem has been solved or does not exist. Conceptual failure involves failure to imagine or speculate about unorthodox course of action. Here the impending threat is from none else than the environmental degradation. The problem exists and has already menaced our very survival on the earth, but the policymakers, and leaders do not recognize it as a crisis of a such a magnitude to deserve special attention and to be kept on the top of their concerns. The environmental crisis is as important as the concern over the 'cognitive dissonance' that distracts the global community and its leadership from the central problem of our time. The focus of the military establishment, the gaze of the media and attention of the public opinion is on sectoral and trivial pursuits. Meanwhile the planet has already slipped into an environmental cul de sac finding no way-out. From the very dawn of human history to the present a major chunk of attention has been on military-political conflicts, territorial or religious disputes quite losing sight of the greater and longer term threat that environmental degradation poses to the very survival of humanity on the surface of the earth. What it requires of the policy makers, world leaders and people is to recognize the scale of the crisis first and then keep it on the top of the agenda so as to devise new political thinking and means of coping with it more quickly and effectively. The Earth Summit of 1992 was a milestone in that it resulted not only in centralizing the international community's attention on the problems of global sustainability and environment but in canonizing what sanctified at Rio as a philosophical lodestar for future guidance and institutionalized policy making - the concept of sustainable development-the ability of the current generation to meet its needs without jeopardizing the ability of the future generations to meet theirs. Agenda 21 scripted out in the summit the need for more research to better understand the carrying capacity of Earth as conditioned by its natural systems, and to develop and apply new analytical and predictive tools in order to assess more accurately the ways in which Earth's natural systems are increasingly influenced by human activity. The character and pattern of development should be refined in a more comprehensive way in the overall context of the capacity of global life support systems to meet the needs of human and non-human life forms.

One of the most useful analytical tool devised in response is ecological footprint that should be embraced by all citizenry and political leadership around the globe.[1] The "ecological footprint" depicts the amount of the planet's biological productivity that an individual, a city, a nation-state or humanity occupies at any one time. It measures the area of biologically productivity land and water required to produce the resources consumed and assimilate the wastes generated, using prevailing technology. The footprint is measured in area units. An area unit equates to one hectare of biologically productive space with average productivity for the world. Thus, 1.0 area unit is equivalent to 0.3 hectare (ha.) of cropland, 0.6 ha of forest, 2.7 ha of grazing land, or 16.3 ha. of coastal sea.[2] It is on the basis of this conceptual development that national and global measurements have been made that carry implications for future generations and challenges for the current political leadership.

In 1996, the earth's biosphere had 12.6 billion ha. of biologically productive space, comprising 9.4 billion ha. of land (49 per cent grazing land, 35 per cent forest, 14 percent cropland, and 2 per cent build up land)

and 3.2 billion ha. of ocean fishing ground. In 1996,with 5.7 billion humans on the planet, this equated to 2.18 hectares per person. With a worldwide population of 6.3 billion today, this average earth share has shrunk to 2.1 ha. per person over the past five years.

Ecological overshooting

In reality, however humanity's footprint is larger than this, at 2.85 units per person. The existing ecological draw-down on the planet exceeds its biologically productive space : the area required to produce food and wood, to give room for infrastructure, and to absorb the CO_2 emissions associated with energy use was 30.7 percent larger than the area available. This excess known as ecological overshooting is the measure by which the current generation is despoliating the planet's natural asset base, drawing on ecological credit from our children's generation reducing the ability of future generation to meet their own legitimate needs.

Over Consumption: Morally and Ethically Indefensible

The findings of the latest study Living Planet Reports 2000 measuring 151 nation states some three quarters of the UN Member States, and almost all of humanity, are notewoththy.

- 75 countries exceeded the "earth share", while 76 countries are below it;
- The 75 countries above earth share account for 21% of humanity;
- The three highest over consumers are the United Arab Emirates (at 16 units), and Singapore and the United States at 12 each or more than five times higher than the earth shares.
- The three modest consumers are Bangladesh, Afghanistan and Eritrea at about 0.5, 0.5 and 0.2 units respectively;
- Three countries living at earth share are Ecuador, Azerbaijan and Cuba.[3]

This shows, the majority of humanity is living below earth share, a minority, i.e., mere 21 per cent of the humanity is living far above it and is responsible for the global "ecological overshoot" that constitutes the threat not only to the present generation but also to the future generations. Most countries are running a national ecological deficit.

Questions of Ethics and Morality

This over consuming minority of humanity is responsible for disastrous consequences of ecological overshooting depriving the future generations of their right to live on this earth. Is their inordinate consumption morally and ethically defensible? Where is the public outcry over this morally indefensible and potentially disastrous overconsumption of the planet by humanity's current generation? Are there any pressures and demands by the countries living below the earth share on these over consuming nations to rethink their pattern of development or put a cap on their level of consumption? Has there been the emergence of any institutional arrangements and consensus to ensure sustainable consumption keeping chime with sustainable development in the North? Are the people in North ready to come down to the level of Cubans or Bangladeshis? These are some of the nagging questions answers to which remain crucial to the solution of the yawning inequity not only between and among nations but also between the present and future generations. As a nation or people we may boast of being a greatest democracy or champion of human rights and torch bearer of high moral values but the national analysis of the "ecological footprint" go a long way to oppugning us for our mistaken and culpable moral offence. It is ironic and dubious that the most profligate and extravagant go scot-free and sermonize on global governance of the environment by hinging on "the meaningful participation" of the developing world.

For example, the only way to stop the doomsday prophecies associated with climate change from coming true, according to scientists is to reduce emission by 50-70 per cent below 1990 levels, by reducing or even altogether abandoning the use of carbon based fuels. The use of these fuels, and hence carbon emissions, are closely linked to economic growth and life style. The more prosperous a country's economy, and the higher its per capita income, the higher its fossil fuel consumption for power generation and transport, and the higher its Greenhouse gas emissions. Industrialized countries owe their current prosperity to years of historical emissions which have accumulated in the atmosphere since the start of industrial revolution and also to a high level of current emissions. Developing countries are very much nascent on the path of industrialization, and their per capita emissions are still comparatively low.[4]

Under these circumstances, any limit on carbon emissions amount to a limit on economic growth turning climate change mitigation into an intensely political issue. International negotiations under the UN framework Convention on Climate Change have turned into a tug of war with rich industrialized unwilling to compromise their life styles and poor developing countries unwilling to accept a premature and inequitable cap on their right to development. The atmosphere is common property resource to which every human being has an equal right. The people of industrialized countries have more than used up their share of absorptive capacity of this atmosphere through their high emission levels in the past and in the present. To that extent, the global warming is their creation. So, it is morally right and binding on them that they take the initial responsibility of reducing emissions while allowing developing countries to achieve at least a basic level of development. Moreover asking developing countries to reduce carbon emissions levels now amounts to denying them their very right to development and freezing their standard of living at their current stage of development. This means some countries will remain more developed than others reinforcing an unequal world rather than mitigating it. Thus, as a result of these crucial questions, the negotiations have turned into a game between two unequal partners.

On the ground that the American life style is not subject to negotiation the Kyoto Protocol aiming at reducing CO_2 emissions agreed to by US President could not be ratified by American Senate. The powerful and dominant economic actors will certainly for the reasons cited above balk at any kind of international negotiation that will challenge the very foundation of their world economy- the carbon energy based technological systems- and the cost of technological transformation will be high and will threaten two of the world's biggest economic genie-the coal industry and petroleum industry. Thus, the resistance from powerful economic actors to any negotiation or policy of rich industrialized nations to compromise their inordinate consumption and life style will be very high, the ecological costs for the rich and powerful will be low, as they have the necessary scientific, technological and material wherewithal to cope with the crisis. On the other hand, the South-countries will face the greatest economic damage resulting from such environmental crisis.

Thus, the question of returning equitable earth share to the deprived majority South countries involves challenging the basis of world economic system, the inordinate consumption and life style of rich North people. Negotiations on this are not merely a matter of economics, but raise high moral concerns for equity, justice and right to life. All future negotiations or talks on sustainable development and environment must hold these moral and ethical concerns as pivotal.

Basudha not Synonymous with World

Liberty and happiness are private pursuits, justifiably behind high economic growth and inordinate consumption and life style. Does the enjoyment of liberty and pursuit of happiness contribute to the common good and are these pursuits at the cost of the others? This world is not a mere conglomeration of disparate atomized dispensable and " at the cost of" individuals and nations knowing no compunction or moral qualms at consuming what comes the way without showing the concerns for others. The beautiful verse from Indian Upanishads, "Basudhaiba Kutumbakam" connotes world is not "Basudha". Basudha is not mere comprising of individuals and nations it comprises both of living and non-living, living together as a cosmic family in unity and harmony. The concept "Basudha" is not mere a conposition and store house of raw materials but a way or an art or living together in an integrative way like the members of a family enjoying with equal rights, authority and dignity. Does the pursuit of liberty and happiness contribute to the harmony, unity and integrative way of "Basudha"? How do they stand in relation to the philosophy behind "Basudha"? Basudha connotes also an infinitude of relations carried into the eternal web of time not a temporal time and space as is mistakenly today held synonymous with mere exploitations and being expendable commodity. Then, liberty and happiness must answer to the philosophical connotation of "Basudha" as well to the future generations—both living and nonliving forming 'Basudha's cosmic relationship in its eternal time frame. Thus disregarding intergenerational principal is a moral inequity of historic proportions.[5]

Economic Rethinking

The crucial question remains whether the people in the North will awaken themselves to the urges of morality for moderating their inordinate life style. So that the lost earth share shall be restored to the deprived. Or the policymakers in North countries realizing their moral responsibility for what so far they have done to the earth will reinvent their economic thinking to ensure a sustainable living for all, making the bond between and among the present generations and also between the present and future generations worth relishing. No signs of either the people in North being awakened to these urges of morality or their governments being serious to rethink the basic foundation of their world economic edifice are perceived today, since the environment is rapidly heading towards a downward spiral.

The illusive link between happiness and economic growth is debatable. The happiness promoted by economic growth, consumerism and wealth production at the private level is not certainly Rousseau's social dimension of enjoying liberty, or J.S.Mill's dissatisfied Socrates' happiness or Maslow's self actualization or transcendence. The happiness derived from inordinate consumption, wealth production and materialistic life style has ruined the earth, deprived the future generation's right to life and pauperized more than half of the globe. A prominent psychologist Edward Diener pointed out that while poverty certainly causes unhappiness, wealth does not necessarily bring happiness. Since 1957,the average American's real income has trebled. But the fraction of Americans who describe themselves as very happy has remained static at one third throughout this period. One reason is because as economist Juliet Schor likes to point out, materialism causes people to spend rather than save, embedding anxiety in daily life. Besides, run away consumerism puts the individual onto a never ending over accelerating treadmill aptly summarized by David Myers' wry comment, "The victor belongs to the spoils". Such a wealth driven happiness is ethically vacuous- not impelled by basic values, and not anchored in concepts of rights and responsibilities. "Not every nation with a high per capita income is truely developed".[6]

To persist in such consumerist happiness by the rich and affluent means to condemn permanently the poor and the future generation to constant penury. Economy has dominated the entire societal space. Adam Smith using the term market helped the transformation of community based subsistence economy into a nationwide political economy covering the entire society and space. The economy tended to overshadow every other reality. Society was understood in terms of nothing else than its functional achievements leaving aside the rest as just folklore or private affairs. The laws of the economy try to dominate the society and not the rules of the society, the economy. The term 'market economy reduces all the societies to mere an 'economy' making the develop mentalists oblivious of the very basic thing that every society has an economy. The market economy as promoted by the west edges its way into all the developmental thinking of the entire world and assumed a deterministic character imprinting its rules and rhythms upon the society and even dictating the drama and the role in it of individuals. The entire globe including all the nations have been reduced to an economy controlled by the west.

Thus, according to the nuances of market economy, development has been identified as economic growth, inordinate consumption and increasing commodity based wants and accumulation of wealth. Poverty is understood in terms of what other people lack or do not have in relation to the west. Destitution and imitating the profligate consumption style of the west, has been the price, the South pays for having ruined its multifarious cultures of frugality, austerity and modesty.

Transforming the Consumer Man

The western economic thinking has reduced a 'whole man' to an economy or insatiated consumer man ever inclining to satisfy his/her desires in terms of consumption or utility. It has reduced all hierarchy of human needs to a mere inventory of wants which are measurable or commensurable quantitatively and translated into consumption items in terms of utility and entered as inputs into the market mechanisms .On the other hand, needs are incommensurable and as such cannot be simply added. Thus, the economic of utility comes out of a worldview or an image of society where all needs are purported to have been met. The shocking thing about this market economy, which relies on this unrealistic picture, is that it begins at a place where fundamental economic problems remain grotesque and are assumed to have been solved.[7]

Thus, the economic rethinking involves the recognition of the existence of a higher self-repository of all good qualities like frugality, austerity, sacrifice, selflessness, materialistic detachment and humanistic concern for community. Over and above, the ordinary lower self is concerned with selfish materialistic consumption giving a free ride to its bodily-derived pleasures. Thus, an economy which reduces all human activity and even existence and values to a sheer commercialization and utilizable commodities in a market place, has ruined the moral, social and ecological fabric of the society, sealing off permanently the mine of alternatives to this kind of market economy. To quote economic humanists :

> The human enrichment of economics involves the recognition that we have a dual self. The self that we feel we are is at the same time perceived as a divided. Its lower aspect- self interest in its unambiguous interpretation-seeks one's own advantage regardless of the interests or well being of others. This is the self that finds it rational to free ride. In contrast, the higher self is that which identifies with a common humanity and finds its rationality in objectivity, fairness, and the truth. The recognition of the existence of the higher self as well as the lower is necessary for an economics that is responsive to human needs and not merely human fancy driven by dollars. This is an economics that just a commercialization of human activity which by itself eats away at the moral, social and ecological fiber of society.[8]

What need not define the individual is not certainly consumerism, acquisitive individualism, unbridled avarice for wealth production and a particular economic and technological way of life that has spelt a doomsday for the present civilization.[9] Amitai Etzion in his 'The Moral Dimension ' is of the view that apart from rationality, which should not be confined to a mere benefit-cost calculus of maximum self-interest, altruism and moral duty are equal motivators of human behaviour in the economic realm. The intrinsic human values, viz., altruism, selflessness, sacrifice, and community life inherent in many great civilizations and cultures can be well cultivated giving a humanistic face to the current consumeristic market economy. For example, in a traditional Mexican village, the private accumulation of wealth results in social ostracism. Gandhian concept of Sarvodaya and swaraj and village republic envisage an economy of self sufficiency and common good, free from the frenzy of inordinate accumulation and extravagant consumerism found in the richest few at the cost of majority poor and the nature. John Keynes, father of neo-classical economics and the chief architect of the post-world war II financial institutions shaped at Bretton-Woods, proposed a toast to economics and economists saying, you are not the trustees of civilizations but one of the possibilities of civilization.

Adam Smith, the father of classical economics in his 'The theory of Moral Sentiments' assumed that a higher providence ordains that the working of market economy remain within certain moral and ethnical bounds not resulting in total selfishness. Adhering to these ethical values inherent in many cultures and civilizations and restoring them to a central place in the economic thinking can usher in such a rethinking.

Misplaced Faith in Human Ingenuity

The economic optimists are of the view that natural resources are more homogenous than commonly thought, since the right institutions and technologies allow substitutions among resources.[10] As a result, the " reservation of particular resources for later use may contribute little to the welfare of future generations". Human ingenuity will eventually find a boundless source of energy. "Advances in fundamental science have made it possible to take advantage of the uniformity of energy / matter-a uniformity that makes it feasible, without reasonable limits, to escape the quantitative constraints imposed by the character of the earth's crust".[11] Further the optimists are effusive in celebrating the creativity and adaptability that make human beings different from other forms of life.

But it is said on the contrary that the optimists often make the mistake of assuming that an adequate supply of the right kinds of ingenuity is always assured. But supply will be constrained by a number of factors including the brain drain out of many poor societies, limited access to capital, and often-incompetent bureaucracies, corrupt judicial systems and weak states. Most importantly, the supply of ingenuity can be restricted by stresses generated by the very resource crises the ingenuity is needed to solve. The poor developing societies face a widening "ingenuity gap" as their requirement for ingenuity to deal with environmental crisis rises while their supply of ingenuity stagnates or drops.[12]

Constraints on Modern Science

There are human cognitive limits. Humans do not have infinite ability to understand and manage the non-linear, multivariate, and often-chaotic processes of ecological-social systems. The relationship in some of these system are "simply too numerous and complex to be grasped much less controlled, by the human intellect".[13] We may never be able to foresee the manifold consequences of our adaptation and intervention strategies. These cognitive limits are likely to be more serious when capital is in short supply, because individual experts and decision makers have to grapple with a greater load of tasks. It is else argued that human beings 'ability to supply ingenuity is probably impeded by cognitive features of human brain that evolved in ecological circumstances entirely different from today's.[14]

The escalating cost of research is another constraint on science. Scientific research generally becomes more costly as it probes further into nature. The poor developing countries cannot afford the high demands for advanced science. For instance, on the issue of tracking climate change, predicting the adverse effects with some degree of reliability and pinpointing responsibility require a degree of investment and scientific expertise that is available mostly to industrialized North. This leaves developing countries dependent on Northern scientists to tell them the extent and fallouts of climate change. Science has been used several times in the past to implicate developing countries, either by showing their future GHG contributions as increasing and counterproductive to industrialized country action or by making no distinction between the 'survival emissions' of the South and the 'luxury emissions' of the North.[15] Very little research has been conducted on the possible impacts of climate change on different countries and region, leaving them totally unprepared to handle the adverse effects of climate change.

The insights of Julian Simon and his band of eco optimists that societies will always be able to summon what is needed to turn environmental adversity into a happy, environmentally benign future, stand discredited. The claims made by them as in the past technological fixes applied by an ingenious species will overcome any global resource constraint, are belied by the fact that unlike in the past, the planet's natural resource base is now deteriorating rapidly. Nowhere is seen the ingenuity to find out alternatives and show the way to stop slide in environmental deterioration.

Recasting National Interest

The politico-military construct of national interest in terms of enhanced power on the realist paradigm in a supposed anarchical international society which is without a sovereign authority led the nation states to be solely concerned with their own interests and survival by any means even at the cost of nature, other nations, and humanity. In their definition of national interest no reference is found to the humanity as a whole or global concerns. In this definition of national interest a dichotomy between 'self-help vs. the other' remained at the center stage. Self help means search for dominance and enhanced military powers and prowess which would enable one to have the world, and the nature at its beck and call. What has pauperized the world and destroyed the nature is this national interest defined by the western nations on realist parlance to meet the inordinate consumptive predilections and luxurious life style of their people- the very basis on which their governments have sought legitimacy from their people. This realist definition of national interest has deemed nature as the spiritless other to be exploited to satisfy the human greed, and taken out religious and ethical concerns from political and economic thinking. It is time that 'planetary interest' become the principal political concept guiding decision-making on survival issue such as sustainability and environmental integrity. It is time that the sovereign nations pursue a legitimate national interest that is, in each case compatible with the greater good of not only the present generation of the entire globe but also that of the future generations. Instead of negotiations being the consequences of competing and contending claims of all national interests defined purely on realist paradigms, the global objective needs to be identified first and accepted as binding with subsequent negotiations determining what national obligations must emanate from this. It is not the dominant few's self-help national interest that would set the tone for global interests of all, but what so far has been termed as the spiritless other would determine the national interest. This kind of view that planetary interests are paramount and accordingly determine the national interest, has been the focus of explicit pronouncements by a host of international leaders for many decades now. In 1996,Barbara Ward urges that "the most rational way of seeing the whole human race today is to see it as the ship's crew of a single spaceship on which all of us, with a remarkable combination of security and vulnerability, are making our pilgrimage through infinity. Our planet is not much more than the capsule within which we have to live as human beings if we are to survive the vast space voyage upon which we have been engaged for hundred of millennia without yet noticing our conditions. This space voyage is totally precarious .We depend on a little envelope of atmosphere for our survival ... we are a ship's company on a small ship."[16]

Similarly in 1993, the president of a global parliamentarian group. Senator Silvia Hernandez of Mexico, spoke of such a notion." The hallmark of our organization" she said, "is the planetary interest-perceiving the world as a single whole, and responding politically with global solutions that reflect an enlightened national interest, not one that is narrowly conceived and competitively pursued".[17]

In March 1994, UN Secretary –General Boutros Boutros-Ghali contented that, "states have always defined their national interests. Today, states must be prepared to accommodate the concept of common-global or planetary interest".[18]

Had this pure vision, global consciousness, insight into the requirements of a global ethics and awakening into a larger spiritual purpose been properly realized, the objective of policymakers to redefine their national interests traditionally pursued on sectoral and narrow national interest would have been amenable to a planetary transfiguration. If the great powers and countries in the North own the historical responsibility for what have been wrought on the planet earth, in the same vein, it is imperative that they take the exemplary initiative in promoting this approach and ensuring that states duly observe it.

Resurgence of new planetary ethics and awakening spirit

In the early 2001, the UN Secretary –General warned that "our ecological footprints on the Earth are heavier than ever before". If we are to lighten our collective footprint on the Earth, we shall require new modes of thinking, pattern of development and life style well informed by religious and spiritual essence. There is no dearth of ethical values for an integrated environment and sustainable development in all-religious faiths. In Hinduism, the Brihad Aranyaka Upanishad clearly declares that God or soul exists in all animate and inanimate living and nonliving beings.

"Even as a caterpillar... reaches out for another blade of grass and draws it over to it, in the same way the soul, leaving the body and unwisdom behind, reaches out to another body and draws itself over to it".

And even as a worker in gold, taking an old ornament, moulds it into a form newer and fairer, even so the soul leaving the body and unwsidom behind, goes into a form newer and fairer : or of the celestial beings, or the gods of light... or a form of the beings".[19]

This shows all life is composed of souls which are passing from lower to higher forms, and occasionally moving in the opposite direction depending on one's Karma. Thus, the human soul has not always been a human soul. Given that popular Hindu Philosophy ascribes up to 83,lakhs rebirths to each soul, the range of incarnational possibilities is immense. All particular forms of beings are thus linked and ultimately transcended by the Atman as it seeks reunion with the Ultimate. Karan Singh in the Hindu declaration made at Assisi put it most succinctly :

"The Hindu viewpoint of nature ... is permeated by a reverence for life, and an awareness that the great forces of nature, the earth, the sky, the water and fire-as well as various orders of life, including plants and trees, forests and animals are all bound to each other within the great rhythms of nature.[20]

The divine is not exterior to creation but immanent in all creation animate and inanimate. Thus, does all creation become sacred and of equal value.

Similarly Buddhism the philosophy of rebirth and life after death, nonviolence and respect for nature declares unequivocally in the words of DalaiLama:

As a Buddhist I believe in the interdependence of all things, the interrelartiship among the whole spectrum of plant and animal life, including the elements of nature which express themselves as mountains, villages, rivers, sky and sunshine.[21]

Judaism, unlike Hinduism and Buddhism, does not hesitate to state that humanity is at the top of the tree, the final and greatest creation of God. But neither is it deluded into thinking that this gives humanity carte blanche to act as it wishes towards the rest of creation .As summarized in the Jewish Declaration on Nature made at Assisi in 1986:

The way that we exercise our power over the rest of God's creatures, over those who ultimately cannot defend themselves against us, must be

the way of love and compassion. If it is not ... the world will soon be destroyed by us.[22]

Christianity shares with Judaism not just its sacred books but also its basic assumption about the nature of God, creation and humanity's special role within creation. Father Lanfranco Serrini exposed all this very simply at Assisi :

In his faithfulness to what he has made and declared to be very good, God, the Father sends his life giving Spirit to renew humanity and the whole Earth, arousing in human beings a firm hope in the life of the world to come, a world of justice, peace, and harmony.

In Islam Allah cares for and oversees all his creation : There is not an animal that lives on the Earth, nor a being that flies on its wings , but forms part of communities like you. We have omitted nothing from the book and they all shall be gathered to their Lord in the end. (Surah 6 : 39)

Both in the Quran in the Shari ah, the legal codes of Islam, the rights of the natural world are strongly expressed and the abuse of them by humanity condemned out of hand. The Quran says: "He set on the Earth firmly rooted, mountains arising above it and blessed the Earth and provided sustenance for all according to their needs".[23]

To say that all religious ethics have remained largely anthropocentric and indifferent to the fate of the natural world is unfounded. Deriving the essence from all-religions about the symbiotic interrelatedness that exists between earth-human beings and cosmos is what would constitute the basis of environmental integrity and sustainable development. However an enlightened and eclectic approach drawing from all religious faiths the spiritual essence to constitute what Tissa Balasuriya calls a planetary theology, is needed. The notion that Earth must been seen as a living organism - in the sense of a homeostatic equilibrium that demands respect from humanity and other terrestrial life forms-has a spiritual dimension on what was initially conceived as a strictly scientific hypothesis.[24] In other words the living earth has a right to life and that right is the primary moral argument for sustainable life. As one philosopher of science Stephen Toulmin writes :

Instead of viewing the world of nature as onlookers from outside, we now have to understand how our own human life and activities operate

as elements within the world of nature. So we must develop a more coordinated view of the world, embracing both the world of nature and the world of humanity...[25]

All said does not mean that there will be an automatic atavism of an ethically and spiritually enlightened character in human beings, irrespective of their attempts at it. Humanity needs to change on two levels : the philosophical spiritual and the level of political – secular action. As Aldo Leopold has pointed out many decades ago, ethics is the recognition of constraints put on an individual as a member of the community to include soils, waters, plants, and animals or collectively the land.[26] Another authority puts it :

"Concern for the environment has to emanate from the basic human capacity to experience the sacred, the capacity to wonder at the blessings that are still with us, to seek after the mysteries of the cosmic order, and a corresponding modesty of the self and its claims on that order".[27]

What it calls for is an ethical shift in human beings. Releasing themselves from their mundane world which is biologically and physiologically informed framework of mind and going beyond this into the philosophical and cosmic realm where they will perceive the unity in all diversities of existence and diversities in that unity-the universalized self, is what constitute the philosophical and spiritual transformation of human beings. It is not a technological fix, nor a matter of new financial investment or exploration of new ingenuity to adapt to the crisis, it is a realization of the transcendental self and an awareness to see the whole creation as divine and sacred. The unending acquitisitive individualism and consumerism born out of a western defined technological and economic way of life that has snapped our eternal and cosmic ties with the whole creation threatening our very survival and existence of all animate and inanimate, has to be replaced by an enlightened, spiritually advanced and liberated self always remaining joyful perceiving his own self as the whole and whole as the self. The attempts at rediscovering the new enlightened and planetary or cosmic ethics informed self in the human beings should be made at both the individual and collective level.

Spiritual Essence : the Basis of Global Policy Decisions

At the political level what is needed a broad spiritual political consensus imbibing the spiritual essence of all religious faiths that will underpin collective action by the entire humanity. No religious faith will dispute the fact that the transformation of the individual acquisitive, biological and physiological self into a universalized self is the key to the solution of the humanity's crisis. What is needed at national and global level is a determined political leadership to carryout this spiritual – political consensus and to inform all aspects of human life by it. When the spiritual essence and political views are weaved into a beatific, broad, coherent, and consistent tapestry that will be the guiding planetary wisdom behind all actions and policy decisions at the national and global level without any way compromising the secular tenor, humanity will begin the move towards a future permeated with justice and equity for all. In that sense of requirement for new vision, new awareness and new thinking, the world summits at Rio or Johannesburg have achieved nothing. These have seen at best orchestration of attempts at reinforcing North's interest by any means leaving the left out crumbs on the table for the poverty stricken and hungry South. South countries have participated as junior partners. The crisis is not as simple as put and understood by the North that it has the necessary technological and material wherewithal to deal with the crisis. The problem is not as put by North one between the North and South or one between the local and global or between rich and poor. It is more a North's proactive initiative required to foster internal changes in its economic and political edifice while responding to the environmental crisis. It is not a question of transferring money or technology but a fundamental question of transforming what is conventionally understood as development.

To persist in such a dwarfed and fractured vision and understanding of the crisis in terms of its consequences not causes, and solutions in terms of technological and economic or scientific ingenuity, at the Earth summits is to lose sight of the labyrinth into which the entire humanity slides and challenges asking us to have a new vision and outlook. The crisis has reached such a magnitude that we owe a lot and remain answerable to the future generations for what we are doing at present. What Oren Lyons admonished us two decades ago remain important till today:

Take care how you place your moccasins upon the earth, step with care, for the faces of the future generations are looking up from earth waiting their turn for life.

Today belongs to us, tomorrow we will give it to the children, but today is ours. You have the mandate, you have the responsibilities. Take care of your people-not yourselves, your people. [28]

Sustainable development would then not be certainly an adhoc response to the selective consequences of environmental crisis, or a paltry response to the demands for doles for the South poor countries or some technological innovations or devising ideological justification for not practicing solidarity with less developed or insistence of the profligate on the disadvantaged South to work for environmental conservation, but a transformation in the human beings most importantly in the rich and affluent in the form of opening up of 'Third eye'- the transcendental vision and cosmic wisdom. What is missing in the agenda of global summits held so far to the perils of living and non-living is the primacy that is to be accorded to this kind of awakening to planetary wisdom and transcendental vision. This is a primary aspect of environmental security based on sustainable development, transcendental vision and cosmic wisdom and recognition of intergenerational equity and accountability.

References

1. W.E.Rees M.Wackernagel, "Ecological Footprints and Appropriated Carrying Capacity : Measuring the Natural Capital Requirements of the Human Economy", in A. Jannson et al, ed., Investing in Natural Capital : The Ecological Economics of the Human Economy (Washington D.C.:Island Press,1994).

2. Kennedy Graham, "Leadership ethics for the future : Tread lightly on the planet", UNU nexions, April 2002, p. 4.

3. Living Planet Report 2000, quoted in Kennedy Grahham, *op.cit.*

4. Anil Agarwal et al., Global Environmental Negotiations 1 (New Delhi : Centre for Science and Environmental, 1999), p. 18.

5. Kennedy Graham, *op.cit.*

6. Denis Goulet, "The United States : A Case of Anti Development", Motive, January-1970, pp. 6-13.

7. A. Lutz Mark and Lux Kenneth, Humanistic Economics : The New Challenge (New York : The Bootstrap Press,1988), p. 29.

8. *Ibid.*

9. John Barkdull, "Why Environmental Ethics Matters to International Relations", Current History Vol. 99, no. 640, November 2000, p. 306.

10. H.E.Goeller and A.M Weinberg, "The Age of Substitutability", Science 191, no. 4228, 20 February 1976, pp. 683-89.

11. Harold Barnett and Chandler Morse, Scarcity and Growth: The Economics of Natural Resources Availability (Baltimore : Johns Hopkins Press for Resourees for the future, 1963), p. 11.

12. Thomas F.Homer Dixon , Environment, Scarcity, and Violence (Princeton/ New Jersey : Princeton University Press,1999), p. 27.

13. Michael Manning and David Rejeski, " Sustainable Development and Risk : A fit ?" Paper presented at the Conference on Comparative Risk and Priority Setting of Air Pollution Issues, Keystone, Colo. 7-11 June 1993).

14. Robert Ornstein and Paul Ehrlich, New World, New Mind : Moving toward Conscious Evolution (New York : Touchstone, 1989).

15. Anil Agarwal, Sunita Narain and Anju Sharma ed., Global Environmental Negotiations 1 (New Delhi : Centre for Science and Environment, 1999), p. 20.

16. Quoted by Kennedy Graham, *op.cit.*, p. 5.

17. *Ibid.*

18. *Ibid.*

19. Brihad Aranyak Upanishad, quoted in Martin Palmer. " The encounter of religion and conservation", in J. Ronald Engel and John Gibb Engel ed., Ethics of Environment and Development (London : Belhaven Press, 1990), p. 53.

20. Assis : Declarations : Messages on Man and Nature from Buddhism, Christianity, Hinduism, Islam and Judaism, Gland, Swizerland : WWF International 1986.

21. Quoted in Foreword in Martin Palmer and Esther Bisset, Worlds of Difference (Glasgow : Blackie, 1985).

22. Quoted in Martin Palmer, " The encounter of religion and conservation", in J.Ronald Engel et al., *op.cit.*, p. 57.

23. Quoted in Martin Palmer, op.cit., p. 61.

24. Kennedy Graham, UNU nexions, April 2002, p. 6.

25. Stephen Toulmin, The Return to Cosmology : Postmodern Science and the Theology of Nature (Berkeley : University of California Press,1982), pp. 255-56.

26. Aldo Leopard, A Sand Country Almanac and Sketches Here and There (London : Oxford University Press, 1980).

27. Rajni Kothari, " Environment, technology and ethics", in J.Ronald Engel and John Gibb Engel, ed, Ethics of Environment and Development : Global Challenge , International Response (London : Belhaven Press, 1990), p. 33.

28. Oren Lyons, " An Iroquois Perspective" in Christopher Vecsey and Robert W. Venables, eds., American Indian Environments: Ecological Issues in Native American History (New York : Syracuse University Press, 1980).

10

Conclusion

In the preceding analysis, the Toronto group headed by Homer Dixon developed a theory of causal links between severe environmental scarcity and violent conflict citing various examples and evidences from numerous case studies. According to his analysis, scarcity or degradation of renewable resources can contribute to civil violence, including insurgencies and ethnic clashes which he predicted, would occur with more severity and frequency than in the past, as scarcities of renewable resources, such as cropland, freshwater, forests, air and atmosphere worsen in many parts of the world. Except in some cases where scarcity's role is direct and immediate cause of violence, in other cases scarcity's role is indirect and in interaction with political, economic, and other factors generates severe social effects that in turn create violent conflict. These social effects are decreased agricultural production, economic decline, poverty, migration and weakening of state apparatus. But analysts who disagree with Dixon and other group's conclusion, tend to interpret these social effects as principal causes of conflict, thus underestimating the scarcity's principal role in causing conflict. As analyzed above, the environmental degradation or scarcity, the increased demand for these resources, and/or their unequal distribution, often interact and reinforce each other to produce resource capture and ecological marginalization.

Resource capture is depiction of a situation when the degradation and depletion of renewable resources interacts with population growth to encourage powerful groups within a society to shift resource distribution in their favour which intensifies scarcity for poorer and weaker groups in the society. Ecological marginalization occurs when unequal resource access in combination with population growth causes long term migration of people dependent on renewable resources for their livelihood to ecologically fragile regions such as steep upland slopes, areas at risk of desertification, tropical rain forests, and low quality public lands within urban areas. The harmful social effects of environmental scarcity in terms of constrained agricultural and economic productivity, migration, social segmentation and state weakening can cause insurgencies, ethnic clashes and coups de'tat.

Skeptics not countenancing the link between environmental scarcity and conflict are of the view that environmental degradation is never a powerful cause of violent conflict. Political, economic and social factors are much more important contributors to violence .[1] Further they argue, these political, economic, and social factors are actually the ultimate causes of environmental scarcity which is therefore, subordinate to these factors. It never plays an independent role. Consequently, it in no way interests the policy makers concerned about the causes of violence in the world. In response to this, Homer Dixon and his group argue that environmental scarcity produces its effects within extremely complex ecological political system. Further more, environmental scarcity is not sufficient by itself, to cause violence except in certain rare cases where the environmental scarcity of renewable resources such as water and fish has been so severe as to cause direct violence. Otherwise, it always interacts with other political, economic, and social factors. Environmental scarcity's causal role can never be separated from these contextual factors. The relationship between environmental scarcity and contextual factors is interactive. It is often impossible to determine the relative weight or power of environmental scarcity as a cause of violence in specific cases. This does not mean that environmental scarcity is always an unimportant cause. Skeptics underestimate the fact that a large proportion of the world population is fully dependent on local cropland, water and forest resources for their existence.

According to Homer Dixon for these reasons environmental scarcity's role as a cause of violence should not be subordinated to that of political economic and social factors. We should not ignore the fact that environmental scarcity often affects the political and economic character of social systems. For example when it stimulates resource capture by powerful elites, it impels institutional and policy reform. Further environmental scarcity is partly a function of the character of a society's physical environment. If severe damage is done to the environment it can become a permanent source of social stress. Twentieth century stands as a witness to a number of inter-state wars over access to non-renewable resources like oil or mineral. But there have been a few wars over renewables such as cropland, plants, fresh water and fish .The reasons are-

First, states cannot easily or quickly translate renewable resources into assets that significantly increase their power. Second, the countries that are dependent on renewable resources are poor. They lack the capability for aggression.

The renewable resource most likely to stimulate inter-state wars is river water. But this occurs in a narrow set of circumstances : the down stream country must be highly dependent on water for its national-well being .The upstream country must be restricting flow, there must be a history of antagonism between the two countries and most importantly the downstream country must be militarily stronger than the upstream country. According to skeptics, since the conflicts arising out of environmental scarcity are not interstate, these are of little concern to policymakers in developed countries. These are of marginal importance to the national security interests of rich countries.[2]

Yet these conflicts generated partly by environmental scarcity, can nonetheless have serious indirect effects on the international community. The changing nature of the international system, heightened economic interdependence, easier access to information, and marvelous revolution in the field of communication resulting in death of distance, make the previously insignificant regions of interest to policy makers. For example, crisis in Haiti, Somalia and Ethiopia often has created serious foreign policy difficulties for developed countries. Flow of refugees due to environmental scarcity to neighbouring countries has often resulted in interstate ethnic violence drawing the attention of international community. Major civil violence within states can affect external trade relations, cause migration, and produce humanitarian disasters that call upon the military and financial resources of developed countries and international organizations.

Finally, the skeptics refute the finding of Homer Dixon' s group that environmental scarcity in combination with social effects will generate violent conflict. According to them, scarcity often stimulates useful technological and social change to adapt to rising consequences of scarcity.

But in the preceding chapters, it has been shown how poor and developing societies will fail to adapt to crisis because of various constraints. In these societies requirements for ingenuity to respond to scarcity outstrips supply of ingenuity, making it vulnerable to scarcity's harsh social effects. Homer Dixon is of the view that in coming decades, worsening environmental scarcities in many regions will further exaggerate the world's already widening hiatus between rich and poor societies and between the powerful and weak people within these societies.[3] The world has been so much interdependent that the wealthy nations should not assume that they will be able to keep themselves immune from turmoil in societies that do not adapt well to scarcity.

While analysing the causes and consequences of environmental degradation and depletion of resources, Homer Dixon and others have ignored the role of rich industrialized North countries in causing damage to the environment. As seen from their research studies, their attention has been only confined to the developing countries. It seems as if the environmental degradation is the doings of the South countries and the rich North countries have squarely put the blame on the burgeoning population of the poor South countries. What is lost sight of is the western pattern of development, its existing economic thinking and life style which have disproportionately used the resources of the Earth to the perils of others. The world economic structure presided over by the United States and its allies has been built at the cost of the environment and poor South. The disastrous consequences of environmental degradation in terms of global warming, sea level rise and hole in ozone layer, are the result of western pattern of development, economic thinking and high life style. The victims of these are the poor South countries whose contribution to emission of CO_2 and other green house gases is very negligible in comparison to that of the rich North countries led by the United States. The profligate life style, exorbitant consumption and commodious and luxurious living of the rich North People have been at the root of the environmental crisis and a greater threat to the poor people in South. These non-military threats more devastating than the nuclear weapons have been ignored in the analysis of Homer Dixon and other researches theorizing the linkage between environmental scarcity and conflict. Rejection of Kyoto Protocol by the United States is suggestive of the fact that America is not willing to compromise its lifestyle for the greater security of the human kind. The process of globalization and militarisation reveals how the notions of free trade, the free market and global security have in reality been an elaborate myth to justify the utilization of violence and state power on behalf of the vested interests. Tom Reifer is quoted to have indicated that " the present model of development is hopelessly flawed, being based on patterns of unjust consumption, exploitations and exclusion which cannot by definition be generalized to all humanity".[4] As pointed out in the introductory chapter, environmental security is not synonymous with environmental conflict. It is beyond the linkage between environmental scarcity and violence to include the prescriptions for rethinking economic pattern, inordinate life style and exorbitant consumption of the western people. It is not western pattern of development, but sustainable development that addresses the rights of future generations, present generations and also those of the nonliving and living to live a healthy life on this planet on the basis of justice and equity, is to be the basis of environmental security. Homer Dixon and other groups' analysis of environmental conflict lacks prescriptions for the way out of the environmental degradation that has led to violent conflict. Environmental security as pointed out above contains prescriptions in terms of a total overhaul of the western pattern of development, established economic thinking and inordinate and profligate life style and consumptive attitude. Intergenerational accountability and responsibility on the basis of justice and equity., sustainable development- and new ethics and vision form the confluence on which environmental security stands to challenge the established frame of mind about security, and sovereign state system associated with the hegemonic mainstream of international relation.

Environmental security is not an appendage to American concept of security built into its system since the industrial revolution. As understood from the viewpoint of the United States environmental security is to assure it access to resources of the earth with the application of military force. Contrary to what is understood by the US, the environmental security does not mean that the poor South countries keep their environment and nature clean and unpolluted to become the resource pool for the rich industrialized North. It is the domestic imperatives shaped by industrial revolution and liberal market economy in terms of production of wealth, profligate consumption, commodious living and inordinate life style that have been built into American concept of security. The entire planet is to be kept ready to fulfill the resource requirement of the North. American is not willing to cast a second look at these imperatives, rather wants that the entire world change according to the domestic demands of the United States.

As seen in the preceding chapter science and political science share the same ontological foundation during 15th & 16th century when individuals were held in the Newtonian language as particles of motion in the direction of wealth production, inordinate consumption, profligate life style and commodious living. Science helped the individuals since industrial revolution seek these goals as sonnum bonnum of life. The liberal market oriented political system emerged then found in the sovereign state system as the guarantor and promoter of these goals to individuals. These formed the basis on which individuals tended their consent and legitimacy to the rightness of action of the sovereign state system to put the Earth, and its bountiful bonanza at the service of the individuals. The result is the accumulated emission of CO_2 and other green house gases to the atmosphere due to industrial revolution, profligate life style and the inordinate consumption which have plagued the entire humanity in terms of global warming, sea level rise and ozone layer depletion. So climate change due to accumulation of CO_2 by the industrialized nations challenges the legitimacy of sovereign state system and the security system built around it. The legitimacy given to the rightness of sovereign state system is not beyond space and time. It has not taken into consideration the very right of the future generations both living and nonliving to live on this Earth. Intergenerational equity and accountability has been ignored. The very basis and criteria of giving legitimacy to the sovereign state system in terms of wealth production, inordinate consumption and profligate life style are to be reconsidered, if environmental security is to be assured to entire humanity. In this reinvented criteria is involved the interests of future generations and their right to survival. Thus, the conventional security system built around sovereign state system and its rightness of action in terms of liberal market economy to produce wealth and promote commodious living, has come in for a reversal and overhauling. Science which has helped state system put these materialistic pleasures as objectives before the individuals, has become a problem in terms of the disastrous consequences of climate change which also challenges the authority of science to come out with alternatives. The new criteria of legitimacy to the sovereign state system and its political – economic thinking based on recognition of the principle of intergenerational equity and intergenerational accountability will certainly bring out a new configuration of Westphalian sovereign state system and the security system built around it. This would give rise to environmental security to all -the present generations, future generations, the living and non-living removing the causes of environmental violent conflict as studied by Homer Dixon and others.

As studied in the last chapter, the ethical foundation of environmental security is quest for new vision and planetary wisdom. Man is no longer the Hobbes and Locke's possessive individual always in the pursuit and direction of sensual pleasures derived from materialistic aggrandizement. Man as mere a consumer and a cog in the market dominated society is different from man as a whole and complete being. The whole world is a cosmic family in which not only human beings of the present generations, but also of the future, living and non-living are bonafide members. Recognizing each other's right to existence and each other as a part and parcel of the cosmic family will bring about new transformed individuals in whose enlarged self is perceived the existence of all other selfs. To transform the body-mind confined individual into a highly enlightened enlarged self encompassing the entire planet, is what is needed to bring modesty to the rapacious nature of the consumerist individuals. As said, a spiritual and political consensus is needed without compromising the secular foundation of the state. The spiritual essence of all religions about the vision of life, nature, cosmos and planet can be weaved into a beautiful ethical tapestry that will constitute the basis of environmental security. The national interest of the sovereign state system defined in terms of self vs. the other- the external world and others are to be exploited for the growth of the self- is to be replaced by planetary interests. The recognition of the organic bond that exists- between man and nature, the future generations' right to live, the principle of inter generational equity and accountability is what is to be identified as planetary interests. Prioritization of these planetary intests over narrow realist framework of interest forms the ethical basis of environmental security.

Homer Dixon and others have established the linkage between environmental scarcity and violent conflict within the conventional realist military framework associated with the state. The response to the situation arising out of environmental conflict, as pointed out by him is through the military apparatus of state, or violent means adopted by the insurgent groups. Then in what way it is different from the traditional conflict or war so far as the response is concerned? Environmental security as outlined above can address the situation by rethinking the traditional concept of security built around the state. It stands at the confluence of sustainable development or rethinking western pattern of development, recognition of intergenerational equity and responsibility and an ethical resurgence in human beings to recast their inordinate life style, exorbitant consumption. Addressing the causes rather than the symptoms can cure the malaise arising out of environmental scarcity. Environmental security addresses the causes rather than the symptoms. It is beyond the prognostic description of the scenarios of violent conflict that will arise out of environmental scarcity or degradation of resources and is prescriptive about the course of action and, policy needed at the level of individual, society, and decision making structure of state with a renewed vision, wisdom and new planetary ethics to save the Earth from the brinks it is being hurtled towards by the mindless, profligate consumption, consumerist market political economy having initiated the entire world into its thesis of development.

References

1. Ronnie Lipschutz, " Environmental Conflict and Environmental Determinism : The Relative Importance of Social and Natural Factors", in Nils Petter Gleditsch ed. Conflict and the Environment (Dordrecht, The Netherlands : Kluwer, 1997), pp.35-50 ; See also Wenche Hauge and Tanja Ellingsen, Beyond Environmental Scarcity : the Causal Pathways to Conflict". Journal of Peace Research 35, no. 3, 1998, pp. 299-317.

2. Jack Goldstone. " Environmental Scarcity and Violent Conflict : A Debate". Environmental Change and Security Project Report. Issue 2 (Washington. D.C. : Woodrow Wilson Center. 1996). p. 68.

3. Thomas F. Homer Dixon, Environment, Scarcity and Violence (New Jersey/ Princeton : Princeton University Press, 1999).

4. Quoted by Audrey E. Kitagawa "Practical Spirituality" in World Affairs Vol.7, no. I, January –March 2003, p. 106.

Bibliography

Earthscan, Environment and Conflict, Earthscan Briefing Document 40, London. International Institute for Environment and Development, 1984.

Ferguson H.L.ed., Changing Atmosphere : Implications for Global Security (Conference Statement, Toronto, 27-30 June 1988), New York : UN General Assembly Document no. A/c.2/43/2,3 October 1988.

Graeger, Nina and Dan Smith, eds., Environment, Poverty , Conflict . PRIO Report.

Oslo : International Peace Research Institute (PRIO),1994.

Lodgaard, Sverre, and Anders H. Afghanistan's Ornas. eds., The Environment and I International Security , PRIO Report , Oslo : International Peace Research Institute (PRIO), 1992.

Nadik S, The State of the World Population 1990(New York : World Population Fund, 1990).

The United Nations Population Fund, The State of World Population 1992(New York: 1992).

World Bank, World Development Report 1988 (New York : Oxford University Press, 1988).

World Bank, World Development Report 1992 : Development and the Environment (Oxford : Oxford University Press 1992).

World Bank, World Development Report 1987 (New York : Oxford University Press, 1987).

World Bank Development Report 1992(Oxford University Press, 1992).

World Bank Development Report 1995(Oxford University Press, 1995).

World Commission on Environment and Development, Our Common Future, The Brundtland Report (New York/London : Oxford University Press , 1987), p. 290.

Books

Bahro Rudolf. From Red to Green : Interviews with New Left Review (London: Verso, 1984).

Bennett, Olivia. Greenwar : Environment and Conflict . Washington D.C. : Panos Institute 1991.

Boserup A. and R. Neild, ed. The Foundation of Defensive Defence (London: Macmillan, 1990).

Boyce James, Agrarian Impasses in Bengal : Institutional Constraints to Technological Change (Oxford : Oxford University Press , 1987).

Brown Lester R.et al. State of the World 1991, World Watch Institute (India), New Delhi, 1991.

Cantori L.J. and Spiegel S.L.ed. International Politics of Regions : A Comparative Approach (New Jersey : Prentice Hall, 1970).

Dalby Simon.Creating the Second Cold War : The Discourse Politics (London : Pinter 1990).

Deudney Daniel H. and Richard A.Matthew, ed., Contested Grounds : Security and Conflict in the New Environmental Politics (Albany : State University of New York Press, 1999).

Enloe C. Bananas, Beaches and Bases : Making Feminist Sense of International Politics (London : Pandora 1989).

Fischer Dietrich . Nonmilitary Aspects of Security : A System's Approach, United Nations Institute for Disarmament and Research (UNIDIR) (Aldershot : Dartmouth Publishing Company Ltd., 1993).

Freeman Peter H. Large Dams and the Environment : Recommendations for Development Planning : A Report for the UN Water Conference, Mar Del Plats, Argentina, March 1977.

Friedmann John. Empowerment : The Politics of Alternative Development (Cambridge : Blackwell Publishers, 1992).

Galtung Johan. "International Development" in Roger A Coate and Jerely A. Rosati, ed., The Power of Human Needs in World Society (Boulder and London : Lynne Rienner Publishers, 1988).

Gilpin Robert. The Political Economy of International Relations (Princeton/ New Jersey: Princeton University Press, 1987).

George P.Castille and Gilbert Kushner, ed., Persistent Peoples : Cultural Enclaves in Perspective (Tuson : University of Arizona Press, 1981).

Giddens Anthony. The Constitution of Society : Outline of the Theory of Structuration (Cambridge : Polity Press, 1984).

Gleditsch, Nils Petter, ed., Conflict and the Environment. Proceedings of the NATO Advanced Research Workshop on Conflict and the Environment, Bolkesjf, Norway, 11-16 June 1996 (Dordrecht, Netherlands : Kluwer, 1997).

Gleick Peter H. " Environmental Resources and International Security and Politics" in E. H. Arnett, ed. Science and International Security : Responding to a Changing World (Washington D.C. : American Association for the Advancement of Science, 1990), p. 512.

Goldsmith Edward and Nicholas Hildyard, "The Social and Environmental Effects of Large Dams", Vol. I Overview, A Report to European Ecological Action Group (ECOROPA) (Cornwell : Wadebridge Ecological Center, 1984).

Grubb M. Green House Effect : Negotiating Targets (London : Royal Institute of International Affairs, 1989).

Gurr Ted. Why Men Rebel (Princeton : Princeton University Press, 1970).

Harrison Paul. The Third Revolution : Environment, Population and a Sustainable World (London/New York :Tauris and Co.Ltd. 1992).

Hillel D. Out of the Earth : Civilization and the Life of the Soil (New York : The Free Press, 1991).

Hollis H.B.., A.L.Powers and M. Sommers,ed. The Conquest of War : Alternative Strategies for Global Security (Boulder, Co. : Westview Press,1989).

Horowitz Donald. Ethnic Groups in Conflict (Berkeley and Los Angeles : University of California Press, 1985).

Hoyt E. Conserving the Wild Relatives of Crops, International Board for Plant Genetic Resources (IUCN) and World Wide Fund for Nature (WWF), Rome and Gland, 1988.

Jacobson Jodi L. Environmental Refugees : A Yardstick of Habitability. World Watch Paper, No.86, 1988.

Job B.ed. The Security Dilemma : National Security of Third World States (Boulder, Co. : Lynne Rienner, 1992).

Johnston, R.J. A Question of Place : Exploring the Practice of Human Geography (Oxford : Basil/Blackwell, 1991).

Kahn H. On Thermonuclear War, edn. 2 (Princeton/ New Jersey : Princeton University Press , 1961).

Kakonen, Jyrki, ed., Perspectives on Environmental Conflict and International Politics (New York, Printer 1992).

Kaldor M. The Imaginary War (Oxford : BasilBlackwell, 1990).

Kaldor M. and R.Falk,ed. Dealignment : A New Foreign Policy Perspective (Oxford : Basil Balckwell, 1987).

Kononova M.M. "Humns of Virgin and Cultivated Soils" in J.E. Gieseking, ed., Soil Components, Vol.I (Berlin : Springer-Verlag, 1975).

Lemarquand David G. International Rivers : The Politics of Cooperation (University of British Columbia and the Waterloo Research Center, 1977).

Leonard Jeffery, "Overview", Environment and the Poor : Development Strategies for a Common Agenda (New Brunswick, N.J. : Transaction, 1989).

Lewin T.H.Wild Races of the Eastern Frontier of India (Delhi : Mittal Publication, 1984).

Linkgreen Goran et al., ed., Peace Processes in the Third World Report from the 1991 Advanced International Programme Conflict Resolution (Development of Peace and Conflict Research, Uppsala University, Sweden 1991).

Moller B. Common Security and Non-offensive Defense : A Non-realist Perspective (Boulder, Co. : Lynne Rienner, 1992).

Nef J.U. War and Human Progress : An Essay on the Rise of Industrial Civilization (Cambridge, Mass. : Harvard University Press , 1950).

Pearce David. Markandya Anil and Barbier Edward B. Blue Print for a Green Economy (London : Earthscan Publications, 1989).

Peters R. and Lovejoy T.ed., Global Warming and Biological Diversity (New Haven : Yale University Press, 1992).

Peterson V.S.ed. Genedered States : Feminist (re)Visions of International Relations Theory (Boulder, Co: Lynee Rienner, 1992).

Rifkin J.Biospheric Politics : A New Consciousness for a New Century (New York : Crown, 1991)

Romm, Joseph J. Defining National Security : The Nonmilitary Aspects (New York : Council on Foreign Relations Press, 1993).

Wachtel Paul . The Poverty of Affluence (Philadelphia : New Society Publishers, 1989).

Walker R.B.J. and S. Mendlowitz, ed. Contending Sovereignties : Redefining Political Community (Boulder, Co. : Lynne Rienner, 1990).

Westing Arthur H., ed. Global Resources and International Conflict : Environmental Factors in Strategic Policy and Action (Oxford: Oxford University Press, 1986).

Westing Arthur H. Comprehensive Security for the Baltic : An Environmental Approach (London/Newbury Park : Sage, 1989).

Periodicals, Journals and Papers

Bachler, Gunther. Conflict and Cooperation in the Light of Global Human – Ecological Transformation ,No.9,1993, (Bern and Zurich : Swiss Peace Foundation and Centre for Security Studies and Conflict Research)

Boge Volker, " Proposal for an Analytical Framework to Grasp Environmental Conflict", Environment and Conflict Project (ENCOP), Occasional Paper No.1, Center for Security Studies and Conflict Research, Bern/ Zurich, July 1992.

Brock Lothar. "Peace through Parks : The Environment on the Peace Research Agenda", Journal of Peace Research Vol. 28 (40)., 1991.

Brown L.Wolfe E. " Soil Erosion : Quiet Crisis in the World Economy", World Watch Paper 60, 1991,World Watch Institute, Washington D.C.:1991.

Brown L.H. "Regional Collaboration in Resolving Third World Conflicts", Survival (London) Vol. 28, 1986, pp. 208-20.

Brown Lester R. and Jodi Jacobson . "The Future of Urbanization : Facing the Ecological and Economic Constraints", World Watch Paper 77, May 1987.

Bruyninckx Hans. "Environmental Security : An Analysis of the Conceptual Problems Defining the Relationship between Environment and Security", a paper presented at Budapest on the Workshop on Environment and Security, organized by the European Peace Research Association,11-16 November 1993, pp. 1-6.

Buzan Barry. "Environment as a Security Issue" in Paul Painchaud, ed., Geopolitical Perspectives on Environment Security, Cahier du GERPE, No.92-05, University Laval, Quebec, p.24 quoted in Ole Waever, Securitization and Desecuritization Working paper 5, 1993,Centre for Peace and Conflict Research, Copenhagen, 1993, p. 14.

Byers Bruce, "Ecoregions, State Sovereignty and Conflict", Bulletin of Peace Proposals Vol.22 (i), 1991.

Dalby Simon."Ecopolitical Discourse : Environmental Security and Political Geography", Progress in Human Geography Vol. 16, no. 4, 1992.

Dalby Simon. "American Security Discourse : The Persistence of Geopolitics", Political Geography Quarterly Vol. 9, 1990, pp.171-88.

Deudney Daniel. "The Case Against Linking Environmental Degradation and National Security", Millennium Vol.19, No. 3, Winter 1990, pp. 416-476.

Deudney Daniel. "Rethinking the Link between Environment and National Security Issues", Perspectives on War and Peace, Centre for International Cooperation and Security Studies, University of Wisconsin, Madison Vol. 7, No. 2, Spring 1990, p. 12.

Dixon Thomas F.Homer. "Environmental Change and Economic Decline in Developing Countries", International Studies Notes Vol.16, No.1, Winter 1991.

Dixon Thomas F.Homer. " On the Threshold : Environmental Change as Causes of Actue Conflict", International Security Vol. 16, No. 2, Fall 1991.

————— "Environmental Scarcities and Violent Conflict; Evidence from Cases", International Security Vol. 19, no. 1, 1994.

————— Boutwell Jeffery and Rathjens George, "Environmental Change and Violent Conflict", Scientific American, February 1993.

Durning A. Poverty and Environment: Reversing the Downward Spiral", World Watch Paper 92, World Watch Institute, Washington D.C.: 1989.

Falkenmark M. "Water: The Silent Messenger between Cause and Effect in Environmental Problems", Water International (Lausanne), 9(2), 1984.

Falkenmark M. " New Ecological Approach to the Water Cycle: Ticket to the Future", Ambio (Stockholm) Vol. 13, 1984.

Finger Mathias. "The Military, the Nation state and the Environment", The Ecologist Vol. 21, no. 5, 1991.

Galtung Johan. "Violence, Peace and Peace Research", Journal of Peace Research Vol. 8, No. 3, 1969, pp. 167-91.

Ganguly Sumit. "Ethno-religious Conflict in South Asia", Survival Vol.35, no.2.Summer 1993.

Gleick Peter. "Water and Conflict: Fresh Water Resources and International Security", International Security Vol. 18, no. 1, Summer 1993, pp. 79-112.

————— "The Implications of Global Climatic Changes for International Security", Climate Change Vol. 15, 1989, pp. 309-25.

————— "Environment and Security: The Clear Connections", The Bulletin of the Atomic Scientists, April, 1991, p. 18.

Gore Albert. "Strategic Environmental Initiative (SEI)", SAIS Review Vol. 10, 1990, pp. 59-71.

Hall D., Mynick H. and Williams R. " Alternative Roles for Biomass in Coping with Greenhouse Warming", Science and Global Security Vol.2, 1991, pp. 139.

Hare F. Kenneth, "Climate, Drought and Desertification", Nature and Resources Vol. 20, no. 1, January-March 1984.

Johansen R.C. "Real Security is Democratic Security", Alternatives Vol. 16, 1991, pp. 209-42.

Kiessling Kerstin and Nathan Keyfitz. "The Population Debate : Urgency of the Problem", a paper presented at the Population Summit of the World's Scientific Academies held in New Delhi from 24-27October 1993.

Krasner Stephen D. "International Relations Theory and Global Environmental Issues", Paper presented at the workshop, Global Resources and Environment: Arenas for Conflict, Opportunities and Cooperation, University of California, Berkeley /Pacific Institute for Studies in Development, Environment and Security, 15-17 March 1990.

Libiszewski Stephan. What is an Environmental Conflict ? Environment and Conflict Project, Centre for Security Studies and Conflict Research, Switzerland, Berne/Zurich, Occasional Paper no.1, July 1992, pp. 1-16.

Lichbach Mark. "An Evaluation of Does Economic Inequality Breed Political Conflict ? Studies", World Politics Vol. 41, July 1989, pp. 431-70.

Lipschutz Ronnie, and John Holdren . "Crossing Borders : Resource Flows, the Global Environment, and International Security", Bulletin of Peace Proposals 21.no2(1990): 121-33

Lodgaard Sverre. "Environmental Conflict Resolution", Paper presented at the UNEP Meeting on Environmental Conflict Resolution, Nairobi, 30 March 1990.

Lodhi Abdul Charles Tilly. "Urbanization, Crime and Collective Violence in 19th Century", American Journal of Sociology Vol. 79, no. 2, September 1973.

Mathews Jessica Tuchman, "Redefining Security", Foreign Affairs Vol. 68, no. 2, Spring 1989.

Myers Norman. "The Environmental Dimension to Security Issues", The Environmentalist Vol. 6, no. 4, Winter 1986.

———"Environment and Security", Foreign Policy, no. 74, (spring 1989), pp. 23-41.

Norse D. "A New Strategy for Feeding a Crowded Planet", Environment 34, no. 5, pp. 6-11, 1992.

Porter Gareth, "Environmental Security as a National Security Issue", Current History, May 1995.

———"Post Cold War Global Environment and Security", Fletcher Forum of World Affairs14, no. 2 (1990), pp. 332-44.

Renner Michael. "National Security: The Economic and Environmental Dimensions", World Watch paper 89, Washington D.C.: World Watch Institute 1989.

Springer A.L. "Protecting the Environment : A New Focus for the Atlantic Alliance", Proceeding of the Academy of Political Science Vol.38, 1991, pp. 124-39.

Teitelbaum Michael S. " The Population Threat", Foreign Affairs Vol.71, No.5, Winter 1992/1993.

Ullman R.H. "Redefining Security", International Security Vol. 8, 1983, pp. 129-53.

Wayne Cornelius Jr. "Urbanization As an Agent in Latin American Political Instability " The Case of Mexico", American Political Science Review Vol. 63, September 1969.

Wendt Alexander, "The Agent Structure Problem" International Organization. Vol. 41, no. 3, Summer 1987.

Westing A.H. "Environmental Security and its Relation to Ethiopia and Sudan", Ambio Vol. 20, 1991, pp. 168-71.

———"Environmental Security for the Danube Basin", Environmental Conservation Vol. 16, no.4, 1989.

———"Comprehensive Human Security and Ecological Realities", Guest Comment, Environmental Conservation Vol. 16, no. 4, Winter 1989.

———"The Environmental Components of Comprehensive Security", Bulletin of Peace Proposal Vol. 20, 1989, pp. 129-34.

———" The Atmosphere as a Common Heritage of Human Kind : Its Role in Environmental Security", Scientific World (London), 34(4), 1990.

Widgrin J. International Migration and Regional Stability", International Affairs Vol. 66, 1990, pp. 749-66.

Index

E